Yuri V. Pleskov

Solar Energy Conversion

A Photoelectrochemical Approach

English by Prem Kumar Dang

With 89 Illustrations

Springer-Verlag Berlin Heidelberg New York
London Paris· Tokyo Hong Kong

Yuri V. Pleskov

A.N. Frumkin Institute of Electrochemistry
Academy of Sciences of the USSR
Leninsky prospekt 31
117071 Moscow, USSR

Library of Congress Cataloging in Publication Data
Pleskov, ÎU. V. (ÎUriĭ Viktorovich), Solar energy conversion : a photoelectrochemical
approach / Yuri V. Pleskov : [translated into] English by Prem Kumar Dang.
Translated from the Russian. Includes bibliographical references.
ISBN-13: 978-3-642-74960-5   e-ISBN-13: 978-3-642-74958-2
DOI: 10.1007/ 978-3-642-74958-2
1. Solar cells. 2. Semiconductors. 3. Photochemistry. 4. Photoelectricity. I. Title.

© Springer-Verlag Berlin Heidelberg 1990
Softcover reprint of the hardcover 1st edition 1990

2151/3020-543210 – Printed on acid-free paper

Preface

In the past 12-15 years an essentially new trend in electrochemistry has sprung up around the problem of solar energy conversion. Strictly speaking, this is not a purely electrochemical but an interdisciplinary field involving the fields of catalysis, corrosion, chemistry of disperse systems, and others. Nevertheless, electrochemistry, to be more exact, photoelectrochemistry of semiconductors, provides a theoretical basis for new methods of converting light energy into electrical or chemical energy, which, we hope, shall find practical application in the not so distant future. In the past years, this field has been discussed amply and at length in special monographs (e.g., in Ref. [1]). Therefore, in this book the photoelectrochemistry of semiconductors is presented in a concise form (exceptions are only specific problems which have been elucidated incorrectly or have not been covered completely in the literature). In this compact monograph we have abandoned the principle of "self-seclusion": for a more deep insight into the fundamentals of electrochemistry, photoelectrochemistry, and physics of semiconductors the reader shall have to refer to the below-cited manuals, while information on the physicochemical properties of particular semiconductor electrodes can be taken, e.g., from Refs. [2, 3]. In contrast, photoelectrochemical conversion of solar energy is discussed in sufficient detail so as not to simply describe the main principles of action and important systems for electrochemical solar energy conversion, but also to underscore the difficulties encountered in their practical realization. The most probable ways of overcoming these difficulties are also discussed. References are given, wherever expedient, of review but not of original articles.

Between 1975 and 1983 the yearly number of publications on the photoelectrochemical conversion of solar energy increased [4]: 1975: 28; 1976: 59; 1977: 127; 1978: 76; 1979: 97; 1980: 180; 1981: 231; 1982: 298; 1983: 212.

Later on, the activity of researchers in this field did not decrease; new information is being gathered at a very intense rate. At the same time, research has advanced to such a level of understanding that it has become possible to make important generalizations, the need for which was long felt. This was an incentive to write this monograph. The reader must not expect to find a formula for making an effective and operative solar cell. Our task is to give the reader an objective and sufficiently complete insight into the contemporary state-of-the-art and the prospects of this important and interesting field.

About this Book

This is the first book completely devoted to the photoelectrochemical conversion of solar energy. The aim is to systematically present the fundamentals of the method of converting solar energy into electrical and chemical energy in photoelectrochemical cells with semiconductor electrodes, to consider the principles of the operation of photoelectrochemical cells of different types, to review important systems, and critically discuss the prospects of developing photoelectrochemical solar cells which are believed to be less expensive, simple to manufacture, and more universal than the traditional solid-state solar cells.

The book is intended for researchers and engineers working in the field of photochemistry, photoelectrochemistry, and photocatalysis. It may also be useful for senior students and post-graduates specializing in these fields.

The Author

Yuri V. Pleskov, D.Sc., was born in 1933. Since 1955 he has been with the A.N. Frumkin Institute of Electrochemistry, Academy of Sciences of the USSR, which he joined after graduating from Moscow State University. His fields of scientific interest are: photoelectrochemistry, electrochemistry of semiconductors, and solar energy conversion. He has to his credit over 150 scientific publications. Also he has written a number of review articles and the following books: Electrochemistry of Semiconductors (in co-authorship with V.A. Myamlin, 1967); Rotating Disc Electrode (in co-authorship with V. Yu. Filinovsky, 1976); Photoelectrochemistry (in co-authorship with Yu.Ya. Gurevich and Z.A. Rotenberg, 1980); and Semiconductor Photoelectrochemistry (in co-authorship with Yu.Ya. Gurevich, 1986).

Contents

List of Symbols

c	electrolyte concentration	n	concentration of electrons in the conduction band
c_{ox} ; c_{red}	concentration of oxidant (reductant) in solution	$N_{A,D}$	concentration of acceptors (donors) in semiconductors
C	differential capacity	p	concentration of holes in the
C_H	capacity of the Helmholtz layer		valence band
C_{sc}	capacity of the space charge layer in semiconductors	P	power
		P_1	power density of the light flux entering the semiconductor
e	absolute value of electron charge	R	resistance
E	electron energy	T	absolute temperature
E_C	energy of the conduction band bottom	V	voltage
		w	work function
E_g	forbidden bandwidth	x	coordinate
E_V	energy of the valence band top	Y	quantum yield
f	fill factor	α	light absorption coefficient
F	electrochemical potential (Fermi level) of electrons in semiconductors	$\alpha_{n,p}$	transfer coefficient for a cathodic reaction occurring through the conduction band (n), the valence band (p)
$F_{dec,n}$; $F_{dec,p}$	electrochemical potential for semiconductor decomposition reactions with participation of electrons (n); holes (p)	γ	quality factor of a Schottky diode
		$\Delta n, \Delta p$	deviation of electron (n) or hole (p) concentration from the equilibrium value
F_{met}	electrochemical potential (Fermi level) of electrons in metals	$\Delta\psi$	contact potential difference
		ε_0	permittivity of free space
$F_{n,p}$	quasi-Fermi level of electrons (n); holes (p)	ε_{sc}	dielectric permittivity of semiconductor
F_{redox}	electrochemical potential of electrons in a solution containing a redox system	η	efficiency of light energy conversion
G	Gibbs' free energy	η	overvoltage
h	Planck's constant	η_H	overvoltage in the Helmholtz layer
i	current density		
i_n	electron current density	η_{sc}	overvoltage in the space charge layer
i_p	hole current density		
i_{ph}	photocurrent	λ	wavelength of light
$i_{sh.c}$	short-circuit current density of a cell	$\tilde{\mu}$	electrochemical potential
		ν	light frequency
J_0	intensity of a light flux entering the semiconductor	$\tau_{n,p}$	lifetime of electrons (n), holes (p)
		Φ	electric potential
k	Boltzmann constant	Φ_H	potential drop in the Helmholtz layer
L_D	Debye length		
L_{sc}	thickness of the space charge layer in semiconductors	Φ_{sc}	potential drop in the space charge layer in semiconductors

φ	electrode potential	A	acceptor
$\varphi^0_{dec,n}$; $\varphi^0_{dec,p}$	equilibrium potential for semi-	c	cathodic
	conductors decomposition reac-	C	conduction band
	tion with participation of elec-	dip	dipole
	trons (n), holes (p)	D	donor
φ_{ext}	external voltage	H	Helmholtz layer
φ_{fb}	flat band potential	max	maximum
φ_{ph}	photopotential	n	electronic
$\varphi^{o.c}_{ph}$	open-circuit photopotential	o.c	open circuit
φ^0_{redox}	equilibrium potential of a redox	ox	oxidant
	system	p	hole
ψ	Volta potential	ph	photo
		red	reductant
		s	surface

Subscripts and Superscripts

		sc	semiconductor; space charge layer
0	equilibrium	sh.c	short circuit
a	anodic	V	valence band

Introduction

1 Role of Solar Energy in the National Economy

The use of the energy of the Sun, as a scientific and technical problem, attracted concentrated attention in the mid-1970s. Until then episodic attempts were made to use this source of power for economic purposes (here, we do not consider the most ancient method of using solar energy for agriculture). The interest in solar energy as well as in other renewable sources of power was aroused by the so-called oil or, roughly, energy crisis of the 1970s. For a consumer of power in industrially developed countries a sudden ten-fold increase in the price of petrol due to the "oil embargo" of Arabian oil-producing countries, and long queues at gas stations were a shock which for several years formed the public opinion as regards the energy policy of these countries. The response of governments and industries was instant: in particular, they allocated large funds for research and development in the field of renewable energy sources.

In the past 12–15 years the attitude towards the energy problem in the world has not always been the same. The first stage (up to the beginning of the 1980s) is marked by large investments and, as a result, rapid advancement in understanding the physical, chemical, and biological problems linked with the development of different types of solar cells and the evolution of manufacturing techniques. At that time it was thought that the energy crisis would become more profound and the oil price (and the cost of power in general) would increase, i.e., the power situation would remain strained for a long time; this was not the case, though.

A number of drastic measures including direct saving of fuel, introduction of power-saving techniques, utilization of new sources of energy (primarily, atomic energy), and others were taken, first of all, in the industrially developed countries dependent mostly on the import of energy resources, and where energy consumption is approximately proportional to the gross national product (Fig. 1). Thanks to these measures, it had become possible to significantly relieve the critical situation by the beginning of 1980s. The price of oil decreased (although still several times higher than that in the beginning of 1970s). This lead to certain coolness in the public opinion regarding renewable energy sources which are currently thought to be more costly and less convenient in use than traditional fossil fuels.

In this monograph, dedicated to a special topic, we shall not dwell on the problem of using solar energy as a whole, as its discussion can be found in other books, for instance in Refs. [5, 6, 7]. Here, we shall restrict ourselves only to the most common considerations. First of all, the above-mentioned crisis is not an "oil" or

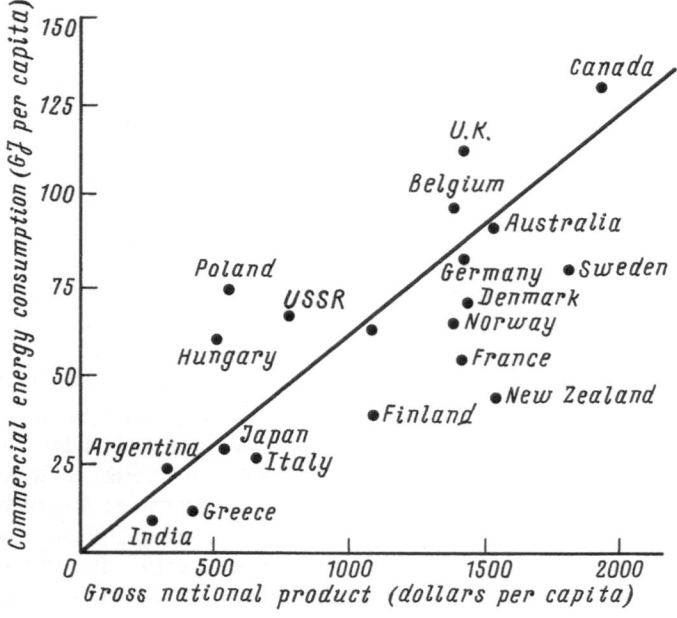

Fig. 1. Per capita energy consumption versus per capita country's gross national product

even a purely energy crisis. In fact, it reflects the crisis in the field of extractive industry as a whole. By now the easily accessible deposits of many mineral resources (not necessarily oil) have been depleted to a more-or-less extent. Extraction has to be carried out in remote areas; comparatively low-grade ores are recovered from a considerable depth. And this, in its turn, involves increased (or ever increasing) expenditure of energy for extraction, transportation, and processing. In agriculture, with the increase in population and the decrease in the reserves of cultivative lands, the use of power-intensive methods has become unavoidable. According to experts' estimates, fossil fuels will suffice only for a limited period; it is believed that the reserves of oil may exhaust somewhere around the year 2000 and those of gas, by the year 2015. (Although coal is expected to suffice for a much longer period.) Besides the purely economic aspect, one must take account of another – ecological – aspect closely related, of course, to the former. Burning of fossil fuels (particularly, coal) appreciably pollutes the environment and may, among other things, give rise to the so-called "greenhouse effect" if the concentration of carbon dioxide in the atmosphere exceeds a certain critical value. Self-release of energy contained in fossil fuels also disturbs the Earth's heat balance; this holds for atomic energy as well. These difficulties will increase with the growth in the consumption of non-renewable energy in the world.

The use of solar energy, on the contrary, will not change the heat balance of the Earth, because its conversion and consumption by man are "built-in" in the natural process of conversion of radiant energy of the Sun into heat; the latter dissipates in space surrounding the Earth. The solar energy conversion process (once the equipment for this has been manufactured and installed) in most of the methods developed for the purpose is "ecologically clean", in particular, waste-free. These main facts permit us to affirm with certainty that solar energy will occupy its place in the energy balance of the future.

2 Photoelectrochemistry as the Theoretical Basis of a Solar Energy Conversion Method

The basis for chemical methods of solar energy conversion is light stimulation of a chemical reaction in the course of which energy[1] is stored (in other words, the reaction products have higher energy than the starting reagents).

Thermochemical methods make use of endothermic reactions such as:

$$CH_4 + H_2O \rightleftarrows CO + 3H_2$$
$$SO_3 \rightleftarrows SO_2 + \tfrac{1}{2}O_2$$

The high temperature necessary for such reactions to proceed is attained by warming the reaction system with solar heat.

In photochemical methods the energy of light is directly used for carrying out the elementary act of chemical reactions. The amount of energy required to overcome the potential barrier of a reaction is provided either in the course of excitation of the reacting particles (molecules, ions) or in the form of electronic excitation of the entire phase (e.g., of the semiconductor) which is a part of the reaction system. Photochemical reactions are usually "quantum" and "threshold" reactions. This means that light energy is absorbed by separate quanta. In this case the quantum energy should exceed a certain value (the threshold) characterizing the potential barrier of the reaction.

Photoelectrochemical reactions are a particular case of photochemical reactions. They proceed at the interface of two conducting phases having different (electronic and ionic) types of conductivity, and, just as electrochemical reactions in general [8], are accompanied by the flow of an electric current in the system.

The development of photoelectrochemistry, as a field of science, in the past decade was stimulated largely by the need for working out a photoelectrochemical solar energy conversion method as a new, ecologically clean and inexhaustible, source of energy. In the most promising variant of the method of converting light energy into electrical and chemical energy, use is made of photoelectrochemical cells with semiconductor electrodes. For better comprehending where semiconductors rank among the objects of photoelectrochemistry, we shall briefly consider different types of photoelectrochemical reactions.

Photoelectrochemistry studies in general the processes involved in mutual

---

[1] Here, one generally talks about enthalpy (heat effect) of chemical substances and more rarely, about free energy (the Gibbs energy).

conversion of light energy and chemical energy in electrochemical systems.[2] Of these, the most common is the conversion of light energy into chemical (or electrical) energy by photoelectrochemical reactions which are accompanied by the appearance of photocurrent in the illuminated electrochemical cell. Less common is the reverse process, i.e., conversion of chemical (or electrical) energy into light energy (manifesting itself as electrochemiluminescence, i.e., emission of light when a current is passed through the cell).

Depending on the area of localization of the "primary" excitation and its nature, all photoelectrochemical processes can be conveniently divided into the following groups:
1. Processes caused by electrode photoexcitation:
    a) metal electrodes
    b) semiconductor electrodes
2. Processes caused by photoexcitation of the electrolyte solution:
    a) reactions of excited ions and molecules
    b) reactions of the solution's photolysis products.

Let us now briefly discuss these processes and their interrelationships. Excitation of an electronic conductor (electrode) under the action of light is caused by the transfer of its valent electrons, after they have absorbed light quanta, to a higher energy level. In a metal, owing to strong interaction in the electron gas, the electron excitation energy dissipates instantaneously, i.e., changes into heat. This significantly restricts the possibility of the development of photoeffects on metal electrodes. Only those excited electrons which have nonzero momentum in the direction normal to the metal surface are able to leave the metal and transfer (in the delocalized state) to the adjoining phase (vacuum, dielectric, electrolyte solution) yielding a photocurrent. This phenomenon is known as photoelectron emission [9].

A semiconductor, as an object of photoelectrochemistry, significantly differs from a metal in that its electron spectrum has a forbidden energy band (a bandgap) that separates the upper of the filled bands of valent electrons (the valence band) from the next empty band (the conduction band). Thanks to the bandgap, the interaction between the electron states in the valence and conduction bands is weak. Therefore, the electrons of the valence band of the semiconductor which, upon excitation by light, had gone into the conduction band and the holes left over in the valence band have a comparatively long lifetime (up to their recombination) sufficient for them to participate in the electrochemical reaction at the electrode/electrolyte interface. It is precisely the photoelectrochemical reactions on semiconductor electrodes, initiated by photogenerated electrons and holes, that will be the topic of this book.

The need for developing a cheap and effective method of converting solar energy into electrical or chemical energy stimulated rapid development of semiconductor electrochemistry in the past decades. The theoretical aspects of this field of

---

[2] In a wider sense, photoelectrochemistry provides a description of various types of changes in the electrode/electrolyte junction characteristics under the action of light in the absence of current as well (for instance, the photopotential or photocapacity at an ideally polarizable semiconductor electrode).

electrochemistry as well as detailed description of photoelectrochemical processes, details of the photoelectrochemical experimental techniques, and a list of the experimentally studied systems can be found in Ref. [1]. In this book we shall focus on the principles of describing photoelectrochemical reactions on semiconductor electrodes, which underlie the action of semiconductor photoelectrochemical solar cells.

In considering the photoelectrochemical reactions caused by the excitation of the electrolyte solution, the properties of individual reagents, say reversible potential different from the potential of the same oxidation-reduction system in the ground (unexcited) state, are usually ascribed to the excited particles, and the laws of electrochemical kinetics developed for dark electrode reactions are applied to them [10]. Note that the lifetime of excited ions and molecules in solution is relatively small, therefore only the substance present in the near-electrode layer, the more so the substance adsorbed on the electrode surface, participates in the considered type of electrode processes. This type of photoelectrochemical processes will be briefly discussed in Sect. 3.5.3 in the context of semiconductor electrode sensitization.

Finally, photoelectrochemistry considers the electrode reactions of stable, long-life final products of homogeneous photochemical reactions which take place in a light-absorbing solution upon illumination. It is precisely these reactions which form the basis of an older method of photoelectrochemical conversion of solar energy into electrical energy in the so-called photogalvanic cells. In a photogalvanic cell, illumination causes photolytic decomposition of the solution with the formation of a mixture of an oxidant and a reductant. With the use of selective electrodes, each being sensitive only either to a reductant or an oxidant, an electrode reaction inverse to the photolytic reaction is carried out in the cell. As a consequence, an electric current flows through the external circuit of the cell, doing useful work. However, the impossibility to fully eliminate side reactions both in the solution bulk and on the electrodes limits the energy conversion efficiency of photogalvanic cells, which is still very low [11].

So far the enumerated fields of photoelectrochemical studies were developed to a large extent independently, and are still very weakly related to each other.

3 Stages of a Photoelectrochemical Process

Every photochemical process starts with the absorption of light quanta, with the result that primary, usually short-lived excited states appear, which contain the energy of the absorbed quantum. They may further disappear in any one of two ways:
a) Upon recombination of the excited states the energy stored in them changes into heat;
b) the stored energy may possibly be given to other components of the system. Here, part of the initially stored energy is spent in vain. Nevertheless, the final (stable) products still retain part of the energy of the absorbed light quantum. This is precisely how the photochemical conversion of light energy takes place.

In order to make pathway (b) more probable than (a), use is made of fast irreversible transformations such as, say, internal rearrangement of the excited molecule (photoisomerization) or electron transitions between the excited particles and their surroundings. Here, only a high rate of such a transformation is not sufficient for attaining the goal, but the transformation should be irreversible and specific, wherever possible. In a homogeneous system, the fulfilment of these requirements causes significant difficulties. That is why increased use of asymmetric (heterogeneous) systems has been made in studies of photochemical energy conversion in the last few years by introducing interfaces of different kinds (in particular, membranes, catalyst particles, etc.) into the reaction scheme.

A semiconductor phase, as an important constituent of a photochemical system, has the following advantages:

a) semiconductors well absorb the electromagnetic radiation, the primary excited states in a semiconductor being excess electrons in the conduction band and positive holes in the valence band;

b) light-generated charges in a semiconductor can be easily separated by applying an electric field. In solid-state semiconductor solar cells with p-n-junctions the electric field is created by the chemical potential gradient of the introduced donor and acceptor impurities; in cells based on the contact of two phases (the so-called Schottky diodes), i.e., semiconductor/metal or semiconductor/electrolyte, the contact potential difference of these phases serve as the source of the electric field.

In the cells wholly made up of electronic conductors (semiconductors with electronic conductivity, metals) light energy is converted only into electrical energy. It is precisely the electrons in the external circuit, where they come at a certain electric potential, that are long-life products at the cell outlet. In cells containing both electronic conductors (metals, semiconductors) and ionic conductors (electrolyte solutions, solid electrolytes) the flow of current will inevitably cause the valent state of substances to change at the interface between the phases having different type of conductivity, i.e., will give rise to electrochemical reactions. Here, light energy may possibly change into chemical energy, too.

4 Advantages of Semiconductor Photoelectrochemical Cells for Solar Energy Conversion

In the past decade the method of converting solar energy with the aid of semiconductor photoelectrochemical cells has been advanced as an alternative to the well-known energy conversion method involving the use of solid-state ("photovoltaic") semiconductor solar cells. The need for this alternative was dictated by two reasons. First, the presently available solar cells are manufactured, as a rule, from highly pure and perfectly crystalline materials, and the p-n-junctions are obtained by using rather sophisticated technology. For this reason, these cells are still very costly, which does not permit their wide application in near-future energetics. In photoelectrochemical cells use is made of an interface which forms on mere dipping the semiconductor into an electrolyte solution, and do not call for the creation of p-n-junctions. Besides, as it has been found in a number of cases, the polycrystalline electrodes made out of not extra-pure materials are noted for their high conversion efficiencies. This is at least partly due to the fact that the semiconductor-liquid phase junction has an off-beat "ideal" structure; there are no mechanical stresses caused by the mismatch of crystal lattices of two contacting solid phases, etc. All this significantly lowers the free carriers recombination velocity at the interface and, as a consequence, raises the conversion efficiency.

Thus, photoelectrochemical cells can in principle be much cheaper than the traditional solid-state cells. This is particularly important because of the comparatively low solar radiation power density requiring the use of large-area converters. The future prospects of the photoelectrochemical solar energy conversion method depends on how completely its potential advantage can be realized in practice.

Secondly, the photoelectrochemical method is convenient in that one of its versions – photoelectrolysis – enables light energy to be directly converted into chemical energy of the photoelectrochemical reaction products, and, thus, permits the energy storage problems to be solved along with energy conversion proper. The need to accumulate energy obviously follows from that the solar radiation power on the Earth's surface strongly depends on daytime, weather, etc., and the maximum of insolation does not coincide in time with the maximum of power consumption.

Chapter 1
Fundamentals of Describing the Structure
of the Semiconductor/Electrolyte Interface
and its Photoelectrochemical Reactions

A semiconductor, as an electrode material, is distinguished by the following salient features:
a) low concentration of free carriers; thanks to this, the external electric field
   penetrates deep into the electrode phase, forming a space charge layer in the
   near-surface region;
b) presence of two types of free carriers – electrons and positive holes, and either
   of these can enter into electrode reactions.

1.1 Aspects of the Physics of Semiconductors

The typical electrophysical properties of a semiconductor – moderate (by comparison, say, with a metal) conductivity and its positive temperature coefficient – are
due to that its electron spectrum has a band of forbidden states (or an energy gap)
between the upper of the filled bands and the next empty band (for greater details
on the physics of semiconductors, see Ref. [12]; in the condensed form it is presented in Chapter 1 of Ref. [1].[1] According to the band theory, a solid body is characterized by a unified electron spectrum in which each electron belongs to the entire body but not to an individual chemical bond. The energy band diagram of
that part of the spectrum which determines the electrical and optical characteristics of a semiconductor in the range of electrical biases and electromagnetic radiation frequencies of interest, is shown in Fig. 2. Here, energy is laid off on the vertical axis; E_V and E_C are respectively the levels of the top of the valence band (upper
of the filled bands) and of the bottom of the conduction band (the next unfilled
band). The difference $E_C - E_V = E_g$ is the width of the forbidden band.

Equilibrium conductivity of a body is ensured by the movement of free carriers that appear due to thermal excitation of valence band electrons. At a temperature higher than absolute zero, $T > 0$, there are always some electrons with kinetic
energy more than E_g, i.e., the energy is sufficient for transferring electrons from
the valence band into the conduction band; in this case, a positive hole remains in
the valence band. In the considered simplest case the equilibrium concentrations
of electrons and holes are equal: $n_0 = p_0$. Both the electrons in the conduction

---

[1] In this chapter, based mainly on the material of Ref. [1] the literature references are given, as a
rule, only to those publications which are not listed in the mentioned book.

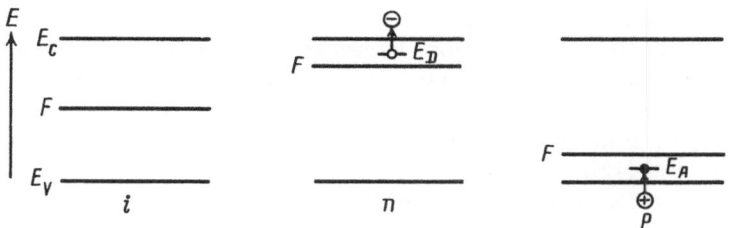

Fig. 2. Energy diagram of semiconductor with intrinsic (i) and n- and p-type impurity conductivity
E_D and E_A are the energies of donors and acceptors, respectively; F - Fermi level

band and the holes in the valence band are involved in the transmission of electrical current through the semiconductor. This is the so-called "intrinsic" conduction.

The electronic conductors with a forbidden bandwidth of up to 3 eV are usually classed with semiconductors, and above 3 eV, with insulators. This classification is very arbitrary, because even at $E_g > 1$ eV the intrinsic electrical conduction is so low that for all practical purposes special dopes creating additional conduction – the so-called "extrinsic", or impurity conduction – have to be introduced. In this case, free carriers appear due to the ionization of impurity atoms: donor-type impurities dissociate with the formation of a (fixed) positive ion in the crystal lattice and of electrons in the conduction band, and the acceptor-type impurities capture a valence electron to give a negatively charged ion and holes in the valence band. Both these processes are schematically shown in Fig. 2 (as one-electron transitions). It might be well to point out that the ionization energy of impurities specially chosen for the purpose, $E_C - E_D$, $E_A - E_V$, amounts to a fraction of a hundredth of an electron volt, i.e., it is less than the average thermal energy of electrons in a crystal at room temperature, $kT/e = 25$ meV. Here, k is the Boltzmann constant and e is the absolute value of electron charge; E_D and E_A are the energies of donors and acceptors, respectively. That is why these impurities get completely ionized even at room temperature; for all practical purposes, $n_0 \simeq N_D$ and $p_0 \simeq N_A$ (N_D and N_A are respectively the concentrations of introduced donors and acceptors).

The concentration of donor or acceptor impurities which are introduced into a semiconductor with the object of imparting it sufficiently high conduction (and, correspondingly, the concentration of conduction electrons or of holes, given by these impurities) is generally taken in the range from 10^{15} to 10^{19} cm^{-3}. The typical donor impurities are: P, Sb, As in silicon and germanium; S, Se, Te in semiconductor $A^{III}B^V$ compounds (such as GaAs, GaP, InP, and others). The typical acceptors are: B, Ga, In in silicon and germanium; Be, Zn, Cd in $A^{III}B^V$ compounds. Crystal structure imperfections as well as vacancies caused due to deviation in the composition of the compound from stoichiometric composition can act as "impurities". For example, in oxides like TiO_2, $SrTiO_3$, Fe_2O_3, and others the oxygen vacancies introduced in a semiconductor in the course of its reductive heating in a hydrogen atmosphere or in a vacuum are donors.

The equilibrium behavior of the electron system of a solid body is described by the electrochemical potential of electrons known as the Fermi level, F. According to this concept, electrochemical potential is the increase in the system's free energy upon adding one electron to it (at constant pressure and temperature).

Now we shall give some quantitative expressions relating n_0 and p_0 to F (Chap. 1 of Ref. [1]; Ref. [12]).

The equilibrium concentrations of electrons and holes are:

$$n_0 = N_C \exp\frac{F - E_c}{kT}; \quad p_0 = N_V \exp\frac{E_V - F}{kT} \tag{1.1}$$

Here, the constants $N_C = 2(2\pi m_C kT/h^2)^{3/2}$ and $N_V = 2(2\pi m_V kT/h^2)^{3/2}$ are called the effective densities-of-state in the conduction and valence bands, respectively; m_C and m_V are the effective masses of electron and hole in these bands; h is the Planck's constant.

The product of concentrations of electrons and holes:

$$n_0 p_0 = n_i^2 = N_C N_V \exp\left(-E_g/kT\right) \tag{1.2}$$

is a constant value (at a given temperature) which, as seen from Eq. (1.2), decreases as the bandgap width increases (n_i is the concentration of free carriers in a semiconductor with "intrinsic" conduction, $n_0 = p_0 = n_i$). Thus, for example, in germanium ($E_g = 0.67$ eV) at T = 300 K, n_i is rather large: 2.5×10^{13} cm^{-3}; in silicon ($E_g = 1.11$ eV) n_i equals only 1.5×10^{10} cm^{-3}.

In a semiconductor with impurity conduction, the product of concentrations of free carriers $n_0 p_0$ is again a constant value (at a given temperature). Because one of these concentrations is strictly given by N_D or N_A (see above) the other is adjusted to it, as demanded by Eq. (1.2). In this case, the concentration of conduction electrons in a n-type semiconductor is much higher than the concentration of holes ($n_0 \simeq N_D \gg p_0$); the electrons thus make a decisive contribution to conduction. Therefore, they are called majority carriers, and the holes, minority carriers. In a p-type semiconductor the majority carriers are holes and the minority carriers are electrons: $p_0 \simeq N_A \gg n_0$.

The Fermi level in a semiconductor having intrinsic conduction lies almost in the middle of the forbidden band:

$$F = \frac{1}{2}(E_V + E_C) - \frac{1}{2}kT \ln\frac{N_C}{N_V} \tag{1.3}$$

(generally, N_C and N_V are of the same order), as is schematically shown in the left-hand part of Fig. 2. In a semiconductor with impurity conduction the Fermi level is shifted towards the edge of majority carriers band. In a n-type semiconductor, the Fermi level lies close to the conduction band bottom; here:

$$F = E_C - kT \ln(N_C/n_0) \tag{1.4}$$

In a p-type semiconductor, the Fermi level lies close to the top of the valence band; in this case:

$$F = E_V + kT \ln (N_V/p_0) \tag{1.5}$$

(cf. right-hand part, Fig. 2).

1.2 Action of Light on Semiconductors

When a semiconductor is exposed to light with energy of quanta more than the electron excitation energy, its electron system comes to a non-equilibrium state. The excitation energy in the most important case of band-to-band transitions is equal to the forbidden bandwidth E_g (or slightly exceeds it). Having absorbed the light quantum, the valence band electron passes into the conduction band and a hole is left in the valence band (Fig. 3, transition 1). In this process, non-equilibrium carriers are generated in pairs such that $\Delta n = \Delta p$. Here, Δn and Δp are excess concentrations of electrons and holes, compared to the equilibrium ones (n_0 and p_0), i.e., $n = n_0 + \Delta n$ and $p = p_0 + \Delta p$, where n and p are the concentrations of electrons and holes in the illuminated semiconductor. This kind of absorption of light is known as intrinsic (or fundamental) absorption.

As light passes through the material of semiconductor, its intensity (i.e., the number of photons per cm^2 per second) decreases as:

$$J = J_0 \exp (-\alpha x) \tag{1.6}$$

Here, x is the coordinate; J_0 is the intensity of light incident on the surface of semiconductor ($x = 0$); α is the linear coefficient of light absorption. The dimension of this coefficient is inverse of length; the quantity α^{-1}, often arbitrarily called the depth of penetration of light into substance, is a measure of distance over which the incident light attenuates e times. The absorption coefficient depends on the radiation frequency ν and microscopic characteristics of the substance. As shown in Fig. 4, with the increase in quantum energy $h\nu$, over a certain threshold value, this coefficient at first increases abruptly and then gradually.

In semiconductors with so-called direct transitions, when the electron energy

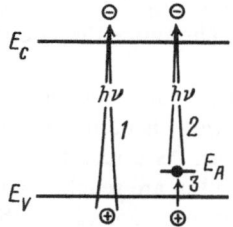

Fig. 3. Light-induced electron transitions in semiconductor due to (1) intrinsic light absorption, (2) impurity light absorption (E_A is the energy of impurity atom which ionizes upon light quantum absorption), (3) "dark" filling of the free level formed in (2) by a valence band electron

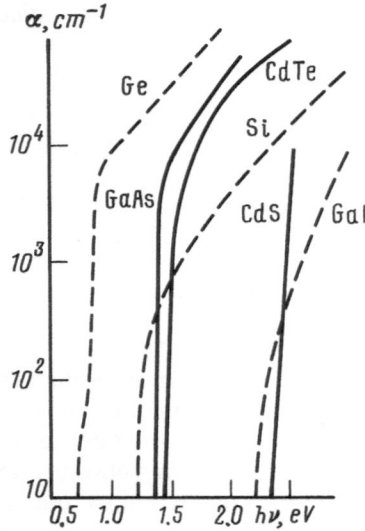

Fig. 4. Dependence of intrinsic light absorption coefficient on quantum energy
Solid lines – materials with direct transitions; dashed lines – materials with indirect transitions

changes without varying its momentum (for greater details, see Sect. 1.3 of Ref. [1]), α increases sharply with the increase in hv and attains larger values, say, for GaAs, CdTe, and CdS. Conversely, in semiconductors with indirect transitions (Si, Ge, and GaP), α varies more smoothly with the increase in hv and has usually a smaller value.

Other light-absorption mechanisms also exist. Of these, we shall mention the so-called impurity absorption. Here, light energy is spent for ionization of atoms of a certain light-absorbing impurity in the semiconductor. As a result, a free carrier appears in any one of the bands (see Fig. 3, transition 2) and the impurity atom acquires a charge of opposite sign and becomes an ion. Further, a free carrier is formed in the other band and "regeneration" of the impurity atom takes place owing to thermal (equilibrium) transition of electron between the formed ion and this band (Fig. 3, transition 3). Of significance is the fact that for the impurity absorption of light the photoexcitation energy is less than the bandgap width (cf. transitions 1 and 2 of Fig. 3). That is why impurity absorption is observed at smaller energies of light quanta (i.e., at larger wavelengths) when intrinsic absorption has not yet occurred. (In Fig. 4 the impurity absorption region (not shown) would lie on the left of the threshold of intrinsic absorption.) Thus, the introduction of light-absorbing impurities sensitizes the semiconductor to more long-wave light. However, it must be borne in mind that light absorption by impurities is relatively poor because of their low concentration. Therefore the coefficient α for impurity absorption of light is small.

Under steady-state conditions, in an illuminated semiconductor the formation of excess (as compared to equilibrium) carriers (photogeneration) is compensated by the inverse process of their disappearance (recombination). As a result, some time-independent concentrations of electrons and holes, n and p, are reached and their product np is no longer equal to $n_0 p_0 = n_i^2$ – a constant which depends on the

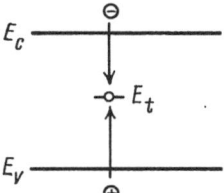

Fig. 5. Schematic of recombination via forbidden band levels E_t is the energy of recombination center

properties of the semiconductor and temperature (see Eq. (1.2)). The values n and p depend on light intensity J_0, absorption coefficient α, and the recombination rate. The latter determines a certain lifetime, τ, of non-equilibrium carriers in a semiconductor. In principle, recombination can proceed in several ways. Of these, the most important one involves the participation of recombination centers with local levels in the forbidden band. Recombination occurs in two steps (Fig. 5): (1) trapping of electron from the conduction band at the unoccupied level; (2) trapping of hole from the valence band at the electron-occupied level. As a result, total energy of the electron-hole pair (equal to about E_g) effectively dissipates in two lots. Thus, the light energy stored during formation of the electron-hole pair changes into heat by recombination, and, therefore, cannot be utilized. The recombination centers may be the atoms of certain impurities, structural defects, and others. In perfect-lattice pure crystals the number of such centers is relatively small. Conversely, in polycrystalline materials (particularly at the grain boundaries) as well as on the semiconductor surface, the recombination rate is usually increased. Intense recombination limits the "light" concentrations of free carriers and is, therefore, a serious handicap for light-energy conversion cells.

1.3 Energy and Electrochemical Potential Levels in the Electrolyte Solution

A solution containing a redox system whose components are related by the equation:

$$Ox + ne^- \rightleftarrows Red \tag{1.7}$$

has a distribution of electron energy levels – filled and empty. The former correspond to the reduced form (Red) with the most probable energy E^0_{red}, and the latter, to the oxidized form (Ox) with the most probable energy E^0_{ox}. Because of fluctuations in the interaction of the dissolved substance with solvent, not a single electron-energy level (E^0_{ox} or E^0_{red}) but a set of levels conforms even to like particles (either Ox or Red) (unlike "monolevel" impurities in a semiconductor where such fluctuations are absent, see Sect. 1.1). The distribution functions of the filled and empty energy levels are schematically shown in Fig. 6 by two bell-shaped lines.

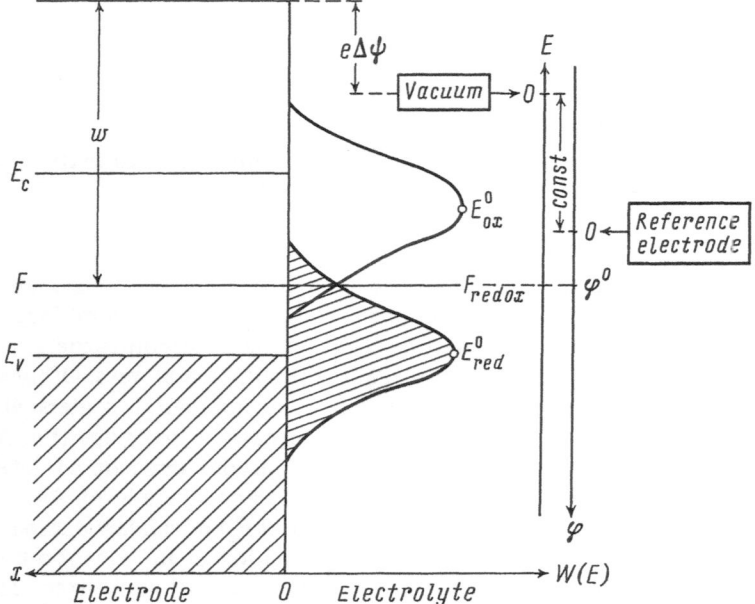

Fig. 6. Energy and electrochemical potential levels in electrode/electrolyte system. The distributions of filled states are hatched

(Within the framework of the most commonly used model, they are described by a Gaussian distribution.) Note that while the effect of fluctuations of polar solvent molecules on the reacting particle energy is shown by two bell-shaped lines in Fig. 6, it is usually presented, according to the theory of the elementary act of electrochemical reactions, as an oscillatory motion along two intersecting energy therms (one conforming to the initial, i.e., reduced state of a particle, and the other, to the final, oxidized state) (see page 184 of Ref. [8]).

An electrochemical potential level may be assigned to the electrons in solution:

$$F_{redox} = F^0 - (kT/n) \ln (c_{ox}/c_{red}) \tag{1.8}$$

where c_{ox} and c_{red} are respectively the concentration of oxidized and reduced particles in the solution; for the simplest model, the constant F^0 is expressed as $F^0 = \frac{1}{2}(E_{ox}^0 + E_{red}^0)$. (In Eq. (1.8), F_{redox} and F^0 are given per particle.)

F_{redox} is often called the "Fermi level of solution" in analogy with the Fermi level of electronic conductors (metal, semiconductor). This term is the source of considerable confusion in certain electrochemical literature. In particular, the "Fermi level of solution" was erroneously called chemical – actually it is the electrochemical – potential of electrons in solution (see, for instance, Ref. [13]). In fact, the solution has no Fermi distribution and no quasi-free electrons but contains only bound electrons in the Red-form and the empty levels in the Ox-form of the redox system.

The electrochemical potential is, of course, an inherent characteristic of the solution phase. Nonetheless, it can be conveniently determined by considering a two-phase solution/electrode system which is at equilibrium due to the occurrence of a redox reaction; see Eq. (1.7). The condition for the system to be at equilibrium is that the electrochemical potentials in the contacting phases be equal:

$$\tilde{\mu}_{red} = \tilde{\mu}_{ox} + n\tilde{\mu}_e \tag{1.9}$$

In this equation, $\tilde{\mu}_{red}$ and $\tilde{\mu}_{ox}$ are the electrochemical potentials of the components of the redox system in solution; $\tilde{\mu}_e$ the electrochemical potential of electrons in the electrode. (It is customary to relate $\tilde{\mu}$ to one mole of respective particles. The relationship between, say, $\tilde{\mu}_e$ and F is expressed as $\tilde{\mu}_e = AF$. Here, A is the Avogadro number.) At the same time the equality:

$$F_{redox} = F \tag{1.10}$$

holds. From this it follows that:

$$F_{redox} = \frac{1}{nA} (\tilde{\mu}_{red} - \tilde{\mu}_{ox}) \tag{1.11}$$

The electrochemical potential level is related to the reversible electrode potential of the same redox system, Eq. (1.7)

$$\varphi^0 = \varphi_0^0 + \frac{kT}{ne} \ln \frac{c_{ox}}{c_{red}} \tag{1.12}$$

(where φ_0^0 is the standard electrode potential) by:

$$F_{redox} = -e\varphi^0 + const \tag{1.13}$$

As will be evident from what follows, the kinetics of electrode reactions can be considered both in terms of reversible electrode potentials and electrochemical potentials of electrons in phases.

Equation (1.13) establishes a relation between the so-called "physical" scale of energy and the "electrochemical" scale of the electrode potential (see Fig. 6). It is customary to refer energy to the energy level of an electron in vacuum, and the electrode potential, to a reference electrode. The value of const appearing in Eq. (1.13) depends on the nature of the solvent and the selected reference electrode. By its physical meaning, this is the free energy of electron transfer from vacuum onto the Fermi level of metal in the reference electrode. (In other words, const equals F_{redox} for the reference electrode.) In part of the electrochemical literature it is known as "absolute electrode potential" (absolute – in the sense that the chosen definition of energy of the electron in the electrode enables the total electromotive force of an electrochemical cell to be divided in a universal manner into two "potentials of a single electrode").[2]

---

[2] A brief and up-to-date presentation of the problem of absolute potential is available in Ref. [14].

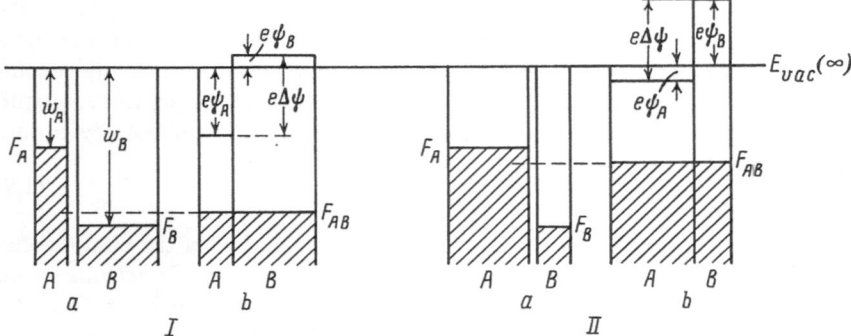

Fig. 7. Schematic showing the dependence of common electrochemical potential level and of outer potentials of two bodies (A and B) in electron equilibrium on their relative dimensions Case I - body B is larger than body A; Case II - body A is larger than body B; a - location prior to contact; b - after the establishment of equilibrium

It is of principle significance from which point in the vacuum the electron energy be counted in the "physical" scale. Let us first, for the sake of simplicity, gain an understanding of this by considering, as an example, the contact of two bodies having electronic conductivity. For the determination of the Fermi level of a separately taken uncharged body, the reference point can be taken arbitrarily because in the absence of an electric field the electron energy at any point in the medium surrounding the body (i.e., in vacuum) is the same. In particular, it is the same both at an infinitely distant point $(E_{vac(\infty)})$ and at the point lying immediately close to the surface of the body (but beyond the effective range of purely surface forces like electrical image forces within the range of the order of 10^{-5} to 10^{-4} cm). The Fermi level of the body relative to the "vacuum level", F_A, is equal to the electronic work function w taken with opposite sign (Fig. 7a). As soon as the body gets charged, say, by coming into contact with the other body (such that electronic equilibrium is established between them, Fig. 7b) a macroscopic electric field appears in the surrounding space. That is why the electron energy is no longer the same at different points of the space. On the other hand, the location of the common Fermi level of two bodies in equilibrium, F_{AB}, cannot be definitely determined. It can be said only that F_{AB} lies between the initial Fermi levels of separately taken bodies, F_A and F_B. The exact value of F_{AB} as well as the outer potentials of these bodies[3], ψ_A and ψ_B, depend on the relative "electron capacity" (or, simply, on relative dimensions) of the bodies. Schematically, this is shown in Fig. 7. The values of ψ can be computed only for the simplest models. Yet their difference $\Delta\psi$ is definite: in absolute value it equals the difference in the work functions $w_A - w_B$ of the bodies divided by the charge of electron. This is the so-called contact potential difference.

Therefore, it is clear that a point in vacuum lying not at infinity (i.e., not that for which $E = E_{vac(\infty)}$) but close to the surface of one of the contacting bodies

---

[3] Outer potential is the difference of potentials between an infinitely distant point, $E_{vac(\infty)}$, and a point near the surface of the body. In the absence of an electric field it equals zero.

should be taken as the reference point of energies (Ref. [15]). If, for example, at a point close to the surface of body B (see Fig. 7) the energy is taken equal to zero, then the level F_{AB} is found from the condition:

$$F_{AB} = -w_B = -w_A - e\,\Delta\psi \qquad (1.14)$$

If one of the bodies has not electronic but ionic conductivity (as, for instance, an electrolyte solution) and the electrons contribute as before to the establishment of equilibrium at the interface (in accord with Eq. (1.7)), then the above-given arguments still hold. It is common practice to take the reference point of energies in the electrode/solution system near the solution's surface because this enables different electrodes to be compared with each other in the same solvent.

As reference point of electrode potentials we take a potential conforming to a certain Red-Ox system (for example, H^+/H_2) for which $\varphi^0 = 0$. For this system, we determine F_{redox} or, which is the same, we find the "absolute potential" const appearing in Eq. (1.13). Since F_{redox} coincides with the Fermi level F of the metal in equilibrium with the solution containing the Red-Ox system (i.e., of the metal which is a constituent of the considered reference electrode) we find, using Eq. (1.14):

$$-F_{redox} = w + e\,\Delta\psi \qquad (1.15)$$

Here, w is the work function for the chosen metal, i.e., the work done by an electron on escaping from the metal into vacuum, and $\Delta\psi$ is the contact potential difference for the metal and the solution of the redox system at its equilibrium potential $\varphi = \varphi^0$ (see Fig. 6).

For example, for an aqueous normal hydrogen electrode (NHE) the "absolute potential" equals -4.43 eV [16]. For the same electrode in solvents other than water the corresponding values of F_{redox} (NHE) have been computed in Ref. [17].

Here it must be stressed that the choice of the reference point in the scale of "absolute potentials" is conditional (but by no means "absolute").[4] In particular, the reference point for different (nonaqueous) solutions does not coincide; each solvent has its own scale. To compare these scales one must know the contact potential difference at the junction between the solvents.

To conclude this section, it must be mentioned that the method of determining the electrochemical potential of a redox solution, based on the use of Eq. (1.15), is a thermodynamic and exact method. Besides, model and, therefore, approximate methods of determining the same quantity (for a comparative evaluation of different methods, see Ref. [18]), have also been proposed. One of the causes of their development is lack of exact information on contact potential differences at the junction between the electrode and the solution, $\Delta\psi$, for many important solvents. If $\Delta\psi$ is not known or cannot be directly measured in any particular system owing to purely experimental reasons, then one may try to estimate F_{redox} by considering potential drops at the phase boundaries in the system. By way of example, we mention a method [19] that involves the use of a composite elec-

---

[4] At the same time it may be mentioned that the so-called "absolute potential" is in fact not a potential but free energy. It is expressed not in volts but in electron volts.

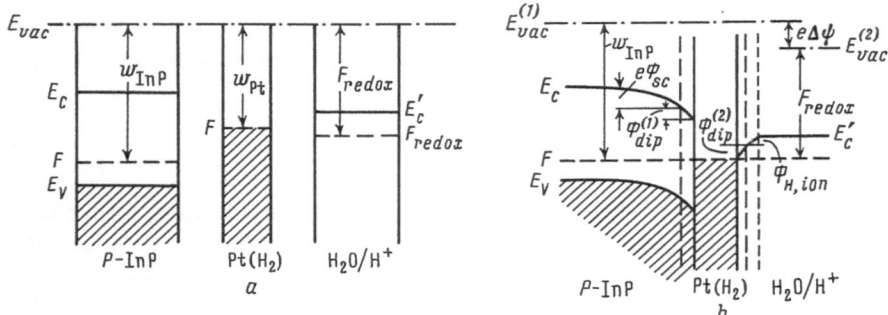

Fig. 8. Schematic of experiment conducted for estimation of "absolute potential" of hydrogen electrode (according to Ref. [19])
Constituents of system before coming into contact (a) and after coming into contact and the establishment of equilibrium (b)
Potential drop in the Helmholtz layer at the Pt/electrolyte solution interface is conditionally shown as two separate components: $\Phi_{H,ion}$ and $\Phi_{dip}^{(2)}$
E_c' is the "conduction band bottom" energy in solution

trode whose substrate is made of a p-type InP semiconductor and which is covered with a layer of hydrogen-saturated platinum (e.g., by cathodic polarization in an aqueous solution). This electrode is kept in an aqueous solution of H⁺ ions of known concentration, thus forming NHE. Figure 8b shows the distribution of the electrical potential in the electrochemical cell obtained by bringing the separately taken uncharged phases (Fig. 8a) into contact in a manner as done in the case shown in Fig. 7. Potential drops occur at the phase boundaries (InP/Pt, Pt/solution) due to mutual charging of the phases when they come into contact. At the InP/Pt boundary these are (a) the potential drop Φ_{sc} in the space charge layer lying wholly in the semiconductor (see next section) and (b) the interphase potential drop proper $\Phi_{dip}^{(1)}$; at the Pt/solution boundary this potential drop includes (c) the ionic component $\Phi_{H,ion}$ as well as (d) the dipole component $\Phi_{dip}^{(2)}$ which owes its appearance primarily to orientated adsorption of water molecules on the electrode surface (for detailed information on the structure of the double layer, see the following section and Ref. [20]). It is not hard to see that the algebraic sum of the enumerated potential drops is equal to the contact potential difference $\Delta\psi$, i.e., to the difference of energy levels of electrons at points close to the free surfaces of InP ($E_{vac}^{(1)}$) and solution ($E_{vac}^{(2)}$). As is shown in Ref. [20], only the potential drop localized within one phase can be measured exactly; in the considered case it is Φ_{sc}. The drop $\Phi_{H,ion}$ caused both by the appearance of an ionic layer and the adsorption of hydrogen on platinum can also be determined. In Ref. [19], Φ_{sc} and $\Phi_{H,ion}$ have been found by separately measuring the differential capacity at the InP/Pt and Pt/solution interfaces. The potential drops $\Phi_{dip}^{(1)}$ and $\Phi_{dip}^{(2)}$ can be but crudely estimated from model concepts. (In Ref. [19] these have been arbitrarily taken equal to zero.) Having estimated $\Delta\psi$ and knowing the work function w_{InP} for InP in vacuum, F_{redox} (NHE) can be roughly calculated from Eq. (1.15).[5]

---

[5] The method mentioned was proposed in Ref. [19] for the aqueous normal hydrogen electrode. There this is an unnecessary method since $\Delta\psi$ for aqueous solutions has been exactly measured in Ref. [18]. Nevertheless, it may prove useful for reference electrodes in nonaqueous solutions.

Returning to Fig. 6 it is seen that the method [19] in its physical sense is based on the estimation of the free energy of electron transfer from the electrode Fermi level to a vacuum level "via solution", and the earlier described method utilizing Eq. (1.15), on the electron transfer "via vacuum".

1.4 Structure of the Semiconductor/Electrolyte Solution Interface

The semiconductor/contacting phase (electrolyte solution, metal or vacuum) interface has typical features which, strictly speaking, determine the photoelectrochemical behavior of the semiconductor and permit the semiconductor to be used for the conversion of light energy.

Generally, there is a charge on the semiconductor surface. On the "free" (i.e., adjacent to vacuum) surface this charge appears due to charging of the so-called surface states (see below); on the semiconductor/condensed phase interface the charge appears also due to the flow of charged particles through the interface. Because of low (compared to metals and even to concentrated electrolyte solutions) concentration of free carriers in the semiconductor (approximate numerical values are listed on p. 10) this charge is not concentrated strictly on the surface of the body, as in the case of metal, but is spread over a certain layer near the surface – the so-called space charge layer of thickness of order:

$$L_D = \left[\frac{\varepsilon_0 \varepsilon_{sc} kT}{e^2 (n_0 + p_0)} \right]^{1/2} \tag{1.16}$$

Here, ε_0 and ε_{sc} are the electrical permittivity of vacuum and semiconductor, respectively. L_D is called the Debye screening length.

In the space charge layer the potential energy of electrons varies with coordinate normal to the interface. In other words, the boundaries of energy bands are bent. The band bending equals $e|\Phi_{sc}|$. Here, Φ_{sc} is the electrostatic potential drop within the charged region. At $\Phi_{sc} > 0$ the bands bend downward and at $\Phi_{sc} < 0$, upward. Here the following cases are possible (see Fig. 9):

accumulation layer: the space charge is composed of majority carriers whose concentration near the surface (n_s, p_s) exceeds their concentration in the bulk of the semiconductor ($n_s > n_0$ or $p_s > p_0$);

inversion layer: the space charge is composed of minority carriers whose concentration near the surface is comparable with the concentration of majority carriers in the bulk;

depletion layer: in this intermediate case (between the above-mentioned ones) the concentration of majority as well as minority carriers near the surface is less than the concentration of majority carriers in the bulk; the space charge is made up mainly of fixed ions of donors (in a n-type semiconductor) or acceptors (in a p-type semiconductor). This is the most common and best known case in photo-

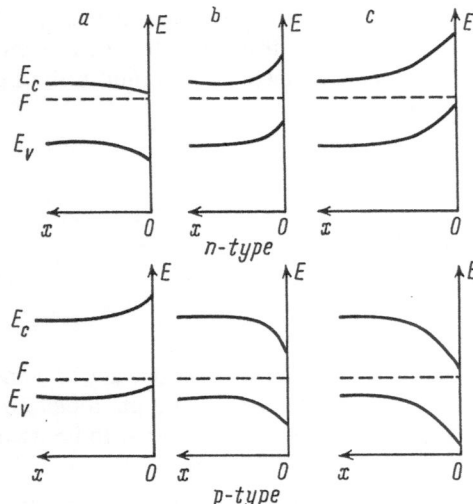

Fig. 9. Schematic representation of band bending for accumulation layer (a), depletion layer (b) and inversion layer (c) for n- and p-type semiconductors

electrochemical conversion of light energy. The depletion layer thickness increases with increase in potential according to the law:

$$L_{sc} = L_D \sqrt{2\,\frac{e\,|\Phi_{sc}|}{kT} - 1} \tag{1.17}$$

where L_D is determined by Eq. (1.16).

In the simplest case the electric double layer at the semiconductor/electrolyte interface is a kind of condenser composed of charges of the space charge layer in the semiconductor (see above) and ions in solution. In concentrated solutions the ionic "plate" consists entirely of ions electrostatically adsorbed on the electrode and localized in the so-called outer Helmholtz plane (Fig. 10a) disposed at distance L_H from the electrode surface, L_H being nearly equal to the radius of solvated ion. Together with the charges on the electrode surface proper the ionic plate forms the so-called dense part of the double layer, or the Helmholtz layer.

In dilute solutions, part of the ionic charge has diffuse structure, as the electronic space charge region in a semiconductor; the thickness of the diffuse (or Gouy) layer is determined by Eq. (1.16) after substituting the dielectric permittivity of the solution for ε_{sc} and the bulk concentration of ions in solution for $(n_0 + p_0)$. A potential drop (ψ_1) analogous to the drop Φ_{sc} occurs in the Gouy layer. Henceforward we shall deal exclusively with moderately concentrated (0.1 to 1 M) solutions in which practically there is no diffuse layer (i.e., there is only the Helmholtz layer at the solution side of the interface), and, therefore, ψ_1 may be neglected.

In a more intricate case which is usually realized in practice, besides the above-enumerated charges, other types of charges enter into the composition of the double layer. First of all, this is the charge of ions specifically adsorbed (chemisorbed) on the electrode. (For describing specific adsorption on semiconductors,

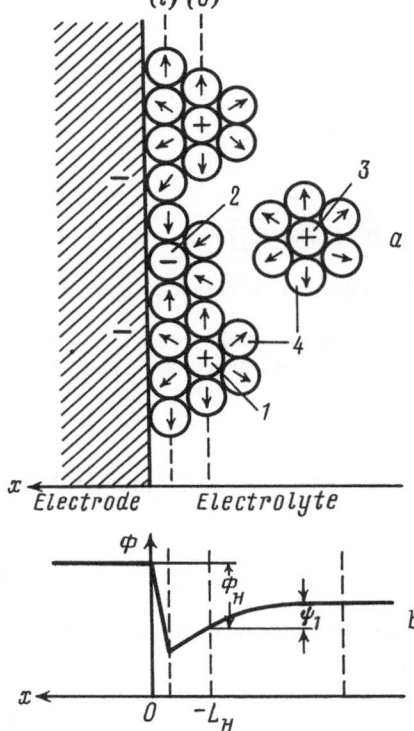

(i) (0)

Electrode   Electrolyte

b

Fig. 10. Structure of electric double layer (a) and distribution of potential in it (b) at the electrode (metal)/electrolyte interface

(i) and (o) stand respectively for inner and outer Helmholtz plane

1 – hydrated cations in the outer Helmholtz plane; 2 – specifically adsorbed anions in the inner Helmholtz plane; 3 – solvated cations in the diffuse ionic layer; 4 – molecules of water

see Ref. [21].) They are located closer to the electrode surface than the electrostatically adsorbed ions and lie in the so-called inner Helmholtz plane (see Fig. 10). Further, in a semiconductor electrode there are charges trapped on the levels lying in the forbidden band of the semiconductor and localized strictly on the semiconductor surface. These, the so-called surface states, appear in part due to the concentration of different kinds of lattice defects and impurity particles on the surface, and other reasons. Sometimes, their number amounts to 10^{13}–10^{14} per cm^2; in this case, the charge trapped in them is comparable in magnitude with the surface charge typical of metal electrodes.

On the whole, the double layer is neutral, i.e., the algebraic sum of all the enumerated charges is equal to zero.

Finally, the orientated dipoles at the interface contributes to the distribution of potential (not charge) in the double layer: these are adsorbed molecules of solvent, polar bonds tying up the surface atoms of the electrode material with chemisorbed particles (often with oxygen atoms), and others. With them is linked the dipole potential drop Φ_{dip} at the interface; its value depends on the number of dipoles in a unit surface area, their degree of orientation and dipole moment. In the subsequent discussion, Φ_{dip} is supposed to be included (if not mentioned otherwise) in the Helmholtz potential drop Φ_H.

In comparing Figs. 9 and 10b it is seen that the total interphase potential drop

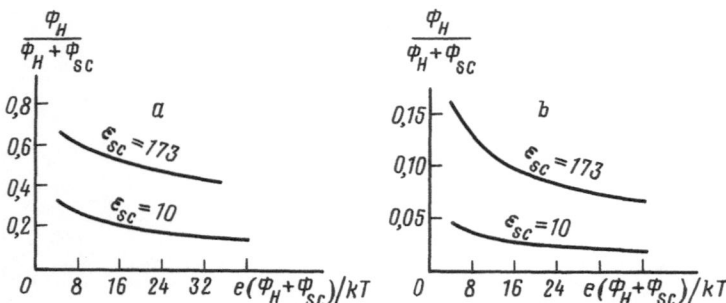

Fig. 11. Computed dependence of the ratio $\Phi_H/(\Phi_H + \Phi_{sc})$ on total dimensionless potential drop $e(\Phi_H + \Phi_{sc})/kT$ [22]
The values of bulk dielectric permittivity of semiconductor are indicated on curves. Bulk concentration of donors (cm^{-3}): a - 10^{19}; b - 10^{17} (Reprinted by the permission of the publisher, The Electrochemical Society, Inc.)

comprises the potential drop in the space charge layer in the semiconductor and the potential drop in the Helmholtz layer[6]:

$$\Phi = \Phi_H - \Phi_{sc} \tag{1.18}$$

In the general case, both these components vary with change in the electrode potential (measured against a certain reference electrode), i.e.:

$$\Delta\Phi = \Delta\Phi_H - \Delta\Phi_{sc} \tag{1.19}$$

The quantitative theory which relates space and surface charges to Φ_{sc} and bulk concentration of free carriers in a semiconductor has been developed in detail (see, for instance, Chap. 3 in Ref. [1]) and well describes the contact phenomena on semiconductors. In particular, the relationships between Φ_H and Φ_{sc} depend on certain conditions, namely on the value of charge trapped on the surface and on the dipole potential drop. In the simplest case, i.e., when there are no specifically adsorbed surface charge and dipoles, from the consideration of electrostatic pattern of the phase contact, it follows that the entire interphase potential drop is practically concentrated in the space charge layer of the semiconductor and the Helmholtz potential drop may be neglected:

$$\Phi \simeq \Phi_{sc}; \quad |\Phi_{sc}| \gg |\Phi_H| \tag{1.20}$$

This is illustrated in Fig. 11 which shows the computed dependence of the ratio of Φ_H to the total interphase potential drop on this potential and dielectric permittivity of the semiconductor. From Fig. 11 b it is seen that at a rather high concentration of the doping impurity ($N_D = 10^{17}$ cm^{-3}) but at a moderate dielectric permit-

---

[6] According to the established tradition, the axes of potentials are directed opposite to each other in reading Φ_H and Φ_{sc}. That is why Φ_H and Φ_{sc} appear in Eq. (1.18) with opposite signs.

tivity of the semiconductor $\varepsilon_{sc} = 10$ (which is typical of many important materials like Si, Ge, GaAs, ZnO, MoSe$_2$, and others) the fraction of Φ_H does not exceed a few percent. It increases with increase in ε_{sc} and N_D (cf. Fig. 11a).

The simplest case is, however, not observed in the experiment because the aforementioned conditions (absence of surface charge and dipoles at the interface) are not fulfilled. On the contrary, on the actual semiconductor surface there is always a considerable amount of the trapped charge and/or surface dipoles. Therefore, the potential drop Φ_H is generally comparable with Φ_{sc} in its absolute value.

At the same time, these components of the interphase potential drop, although they are close to each other in absolute value, may behave differently when the electrode potential is varied. Below we shall dwell on this problem.

The semiconductor electrode potential at which $\Phi_{sc} = 0$ is called the flat band potential (φ_{fb}). As follows from the name and definition of this concept, at $\varphi = \varphi_{fb}$ the bands are not bent (the bands are "flat" right up to the semiconductor surface); this has a decisive effect on the formation of photocurrent (see Sect. 2.1). Therefore, the flat band potential is a critical characteristic of a semiconductor photoelectrode.

The flat band potential is determined by the most common method of measuring the differential capacity. When applied to the widely encountered case of a depletion layer, the space charge theory yields the following relationship between capacity and potential:

$$C_{sc}^{-2} = \frac{2}{\varepsilon_0 \varepsilon_{sc}\, eN_D}\, (|\Phi_{sc}| - kT/e) \qquad (1.21)$$

According to this relationship, one expects to obtain a straight line upon representing the experimentally measured capacity C as a function of the electrode potential φ in $C^{-2} - \varphi$ coordinates (the so-called Mott-Schottky plot, see Fig. 12). Extrapolation of this line up to the intersection with the potential axis yields the flat band potential φ_{fb} accurate to kT/e. Knowing the slope of this line one can find the concentration of the doping impurity (N_D) provided $C = C_{sc}$ and, according to Eq. (1.20), $\Delta\varphi = |\Delta\Phi_{sc}|$ (for detailed consideration of conditions of applicability of Eq. (1.21), see Sect. 3.7 in Ref. [1]).

Now we can construct the energy diagram of the semiconductor/solution con-

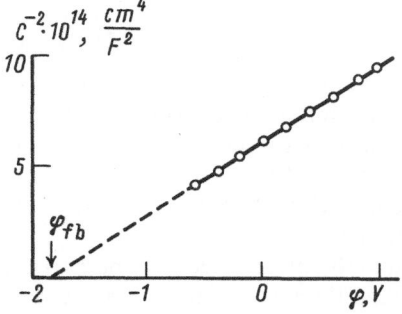

Fig. 12. Mott-Schottky plot for n-type CdTe electrode in 1 M NaOH [23]

Frequency 100 kHz; potentials are given against mercury sulfate electrode (Reprinted by permission of the publisher, The Electrochemical Society, Inc.)

tact, showing the levels of energy and electrochemical potential in the semiconductor phase against a certain reference electrode and/or vacuum. Below we shall often rely on such diagrams for the discussion of photoelectrochemical processes in semiconductor solar cells.

Having determined the flat band potential, we thus find the location of the Fermi level F of the semiconductor in the scale of electrode potentials. Recall that the Fermi levels of two electrodes at the same potential coincide. Using Eq. (1.13) we find the position of F on the energy scale. The location of the majority carriers band edge E_C (or E_V) relative to F is readily found by Eq. (1.4) (or Eq. (1.5)). Finally, knowing the width of the forbidden band $E_g = E_C - E_V$, we ascertain the location of the minority carriers band edge.

As at $\varphi = \varphi_{fb}$, by definition, there is no potential drop in the semiconductor, i.e., $\Phi_{sc} = 0$, the variation in the flat band potential when the external conditions are altered is completely caused by the change in the potential drop in the Helmholtz layer Φ_H.

Let us now see how the composition of solution and the electrode polarization affect the potential drop in the Helmholtz layer.

The dependence of Φ_H on the composition of the solution can be conveniently analyzed by considering as an example the oxide semiconductor MO_n in aqueous solutions of different pH. The metal-and-oxygen-containing groups on the solid-body surface dissociate upon reacting with the solvent according to either the acidic or basic type ("s" stands for surface):

$$(M{=}O)_s + H_2O \; \rightleftarrows \; (M^+{-}OH)_s + OH^-$$

or   (1.22)

$$(M{=}O)_s + H_2O \; \rightleftarrows \; \left(M{\Big\langle}{}^{O^-}_{OH}\right)_s + H^+$$

Here, the dissociation equilibrium depends on the pH of the solution. Dissociation gives rise to a potential drop in the Helmholtz layer, which depends on the surface density of ionized groups. Simple model calculations reveal that this potential drop varies with pH, according to the equation:

$$\Delta\Phi_H = -\frac{2.3kT}{e}(pH)$$ (1.23)

In conformity with this equation, the straight line (the Mott-Schottky plot) is expected to shift parallel to itself along the potential axis under changes of pH. This is what exactly happened in the experiment conducted, e.g., with ZnO (Fig. 13) and other oxide (TiO_2, Fe_2O_3, etc.) electrodes as well as with oxidized semiconductor electrodes (Ge, GaAs, and other materials carrying chemisorbed oxygen on the surface). Such a dependence of Φ_H on the concentration of ions forming an ionic double layer is also observed in the case of a CdS electrode in sulfide solution (S^{2-} ions are chemisorbed on the CdS surface).

Let us now turn to the dependences of Φ_H and Φ_{sc} on the electrode potential φ. As already mentioned, in the simplest case (when there is no charge on the semi-

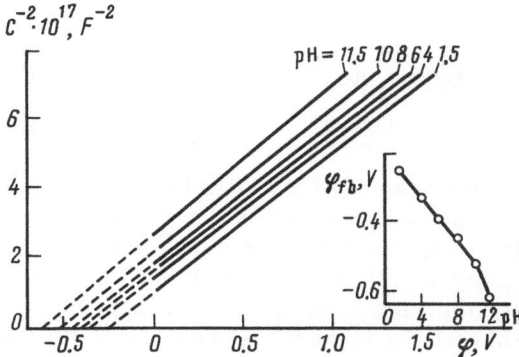

$C^{-2} \cdot 10^{17}, F^{-2}$

pH = 11.5 10 8 6 4 1.5

φ_{fb}, V

Fig. 13. Mott-Schottky plot for ZnO electrode [24]
The values of solution pH are shown on the curves. The inset shows flat band potential versus pH

conductor surface) the following condition $|\Delta\Phi_H| \ll |\Delta\Phi_{sc}|$ should hold. From this condition it follows that $|\Delta\Phi_{sc}| \simeq |\Delta\varphi|$. But in the experiment a more complicated situation is faced. By way of illustration, the $\Phi_{sc} - \varphi$ curve for a germanium electrode is shown in Fig. 14. This curve comprises two types of dependences. When the electrode potential is scanned from negative to positive values, $|d\Phi_{sc}/d\varphi| \simeq 1$ (in the range from -0.6 to -0.4 V), i.e., $\Delta\Phi_H = 0$. Further, between -0.4 and $+0.2$ V the potential drop in the semiconductor, Φ_{sc}, is almost constant, i.e., $\Delta\Phi_H \simeq \Delta\varphi$. Finally, at potentials more positive than $+0.2$ V, $|d\Phi_{sc}/d\varphi| \simeq 1$ and $\Delta\Phi_H = 0$. Thus, in some intermediate region of potentials the entire change in the interphase potential drop is accounted for by the Helmholtz layer; on the contrary, in the extreme regions, the potential in the Helmholtz layer remains unchanged but the entire potential change takes place in the space charge region (the causes of this will be discussed below).

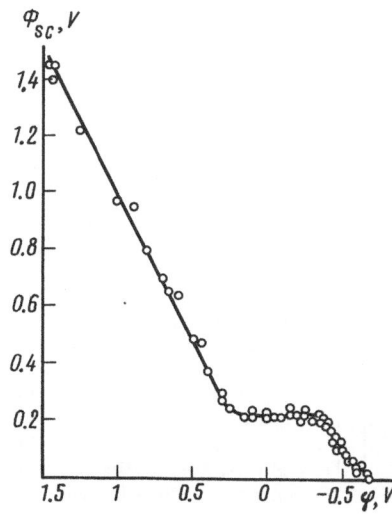

Φ_{sc}, V

Fig. 14. Dependence of potential drop in the space charge layer on electrode potential for germanium in 48 % HF [25]

These two limiting cases:

(1) $|\Delta\Phi_{sc}| \gg |\Delta\Phi_H|$,   i.e., $|\Delta\varphi| \simeq |\Delta\Phi_{sc}|$ (1.24a)

(2) $|\Delta\Phi_{sc}| \ll |\Delta\Phi_H|$,   i.e., $|\Delta\varphi| \simeq |\Delta\Phi_H|$ (1.24b)

are shown on the energy diagram of the contact (Fig. 15). As is seen from this figure, in the former case (Fig. 15a) the energy band edges on the surface, $E_{C,s}$ and $E_{V,s}$, remain unchanged relative to the energy level in solution (and relative to the reference electrode); only the band bending in the semiconductor changes. In the latter case (Fig. 15b), on the contrary, the band bending does not change but all energy levels in the semiconductor shift relative to the levels in the solution. Here, the difference between $E_{C,s}$ (or $E_{V,s}$) and the Fermi level F at the semiconductor surface remains unaltered. In the most recent literature, case (1) is known as "band-edge pinning at the surface", and case (2), as "Fermi-level pinning at the surface".

An intermediate case ("partial Fermi-level pinning") is often observed in the experiments: $\Delta\Phi_H = a\Delta\varphi$, where a takes a value between 0 and 1 (the literature data are reviewed, for instance, in Ref. [26]).

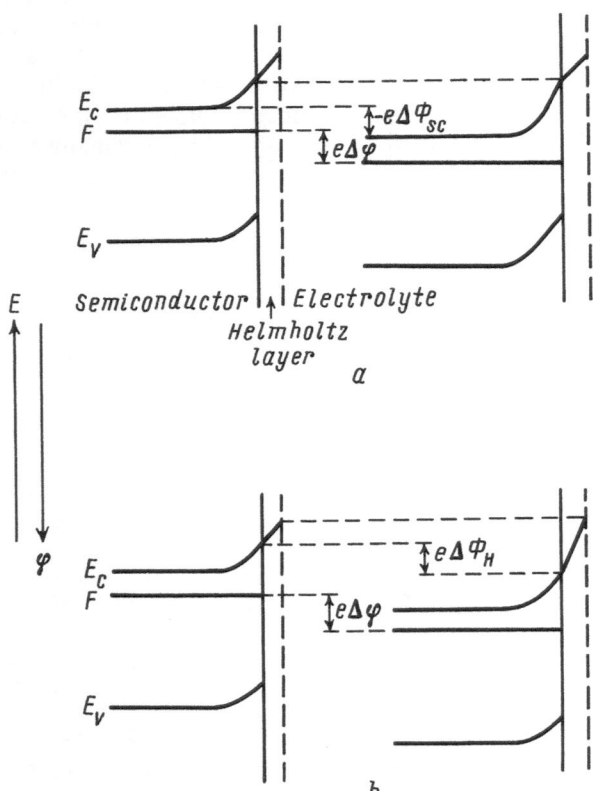

Fig. 15. Energy diagram showing energy band edges pinning (a) and Fermi-level pinning (b) at the semiconductor electrode surface
On going from the left- to the right-hand figure the electrode potential changes by $\Delta\varphi$

The following may be the cause of complete or partial "Fermi-level pinning at the surface", or, which is the same, of large variation of potential drop in the Helmholtz layer:

1. *Large concentration of surface states* whose charge varies in the considered potential ranges. An analysis reveals that a concentration of 10^{13} to 10^{14} cm^{-2} suffices for effective Fermi-level pinning. At a certain energy distribution of these levels, simultaneous variation of Φ_{sc} and Φ_H is possible such that $\Delta\Phi_{sc}/\Delta\Phi_H = $ const; in this case, the Mott-Schottky plot is a straight line (cf. Eq. (1.21)) whose slope is, however no more equal to $2/\varepsilon_0\varepsilon_{sc} eN_D$. Therefore, it cannot be used to compute N_D [27].

2. *Strong charging of semiconductor electrode.* Thus, under strong cathodic polarization of a n-type semiconductor electrode the space charge region is an accumulation layer (cf. Fig. 9a); according to the space charge theory, with the increase in polarization of electrode the variation of potential drop in the accumulation layer slows down and finally stops as the variation of potential drop in the Helmholtz layer, on the contrary, becomes relatively large. A similar relative variation of Φ_{sc} and Φ_H should be observed also in another limiting case – when an inversion layer is formed (Fig. 9c), e.g., under strong anodic polarization of a n-type semiconductor electrode. The question whether the inversion layer can be really formed on semiconductor electrodes under actual conditions, i.e., when current flows through the interface, has not been completely elucidated. In some cases an inversion layer is apparently formed [28]. Nevertheless, as the reaction proceeds with the involvement of minority carriers, the latter are more often extracted from the near-surface layer into solution, their near-surface concentration remains low and therefore the depletion layer is retained (in any case, in darkness) instead of the formation of an inversion layer. To this points, for example, the typical monotonically decreasing dependence of semiconductor surface conductivity on potential (Fig. 16).

3. *Variation of dipole potential drop* at the interface. The chemisorption processes that occur in the course of variation of electrode potential may be accompanied by the formation or disappearance of adsorbate-semiconductor polar bonds such that the dipole component of the Helmholtz potential drop, Φ_{dip}, is affected. Thus, for example, in a certain range of potentials the variation of Φ_H on germanium electrode (see above, Fig. 14) is explained by that the "hydrid" type of surface (germanium is coated with chemisorbed hydrogen) gradually changes into a "hydroxide" type of surface (germanium is covered with

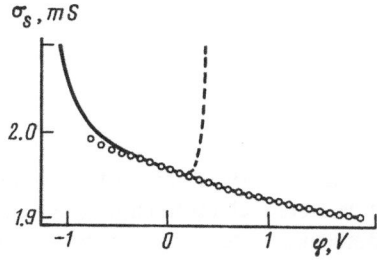

Fig. 16. Dependence of surface conductivity σ_s of n-type GaAs electrode in contact with 0.5 M H$_2$SO$_4$ solution on potential [29]
Solid line - calculated by space charge theory for depletion layer; dashed line - the same taking account of inversion; points - experimental data

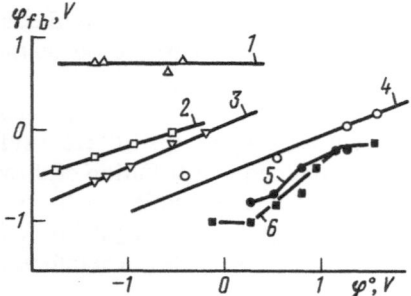

Fig. 17. Dependence of flat band potential on reversible potential of redox system in acetonitrile solutions [30] Electrode material: 1 – p-type WSe_2; 2 – p-type GaAs; 3 – p-type Si; 4 – p-type InP; 5 – CdS; 6 – TiO_2 (Reprinted by the permission of the publisher, The Electrochemical Society, Inc.)

chemisorbed oxygen, more precisely with hydroxyl groups). And these two states of the surface have different total dipole moments.

The described concepts of pinning of band edges or Fermi level are applicable also for the case when instead of applying external bias, the composition of the redox system in solution is changed, i.e., the Fermi level F of the semiconductor electrode is shifted not electrically but chemically. Here, advantage is taken of the variation in the electrochemical potential F_{redox} of the solution in equilibrium with the semiconductor. In this case, the dependence of Φ_H on φ^0 (φ^0 is the reversible potential of the redox system) resembles the above-considered dependence of Φ_H on φ. It can be conveniently plotted as a dependence of flat band potential, φ_{fb}, on φ^0 (Fig. 17). Choosing the composition and concentration of the redox system the reversible potential (and, accordingly, F_{redox}) was varied over a wide range. In the "ideal" case, φ_{fb} should be independent of φ^0 (the case of band edges pinning at the surface, cf. Eq. (1.24a)); this is observed in the experiment conducted with a p-type WSe_2 electrode, indeed. In the opposite limiting case (Fermi-level pinning at the surface, Eq. (1.24b)) φ_{fb} (being a function of Φ_H) depends linearly on φ^0. This is typical of p-type GaAs, Si, and InP electrodes. In case of CdS and TiO_2 electrodes, a change-over from band-edge pinning in one range of electrode potentials to Fermi-level pinning in the other range is observed. A particular behavior depends on the density and energy spectrum of surface states at the semiconductor/solution interface.

The effect of band-edge pinning or Fermi-level pinning shows up also when the electrode potential varies due to illumination of the electrode (see Sect. 2.1).

1.5 Kinetics of Electrochemical Reactions on Semiconductor Electrodes

The distinguishing features of electrochemical reactions on semiconductors depend, first, on that there are two types of current carriers – electrons and holes – in the semiconductor, of which either can participate in charge transfer through the phase boundary. Therefore, Eq. (1.7) should be rewritten in the more general form:

$$Ox + e^- \rightleftarrows Red \qquad (1.25\,a)$$

$$Ox \rightleftarrows Red + h^+ \qquad (1.25\,b)$$

where h^+ stands for a hole in the valence band. Electron transitions take place between the conduction and valence band of the semiconductor, on the one hand, and the filled or free levels in solution, on the other hand. In total four electronic and hole currents passing through the phase boundary correspond to Eq. (1.25); these are shown in Fig. 18 by arrows. According to the widely used model of the elementary act of electrode reactions, electrons transfer between (initial and final) equal-energy levels. Therefore, electrode reactions will occur when the distributions of electron levels overlap each other in both phases. Depending on the value of the reversible potential φ^0 of the redox system, a set of its levels coincides, as shown in Fig. 18, mainly either with the semiconductor conduction band (more negative potentials φ^0) or with the valence band (more positive potentials φ^0). That is why there is more possibility of the occurrence of reduction reactions with the involvement of conduction band electrons, and of oxidation reactions, with the participation of valence band holes.

Second, the concentration of current carriers in semiconductors is low compared to metals and concentrated solutions of electrolytes. This is the reason why the kinetic equation of reaction contains, besides the usual activation term and concentration of the reacting particles (ions or molecules) in solution, the concentration of those electric charges (electrons or holes) in the semiconductor which participate in the reaction. Thus, for example, for the anodic current flowing

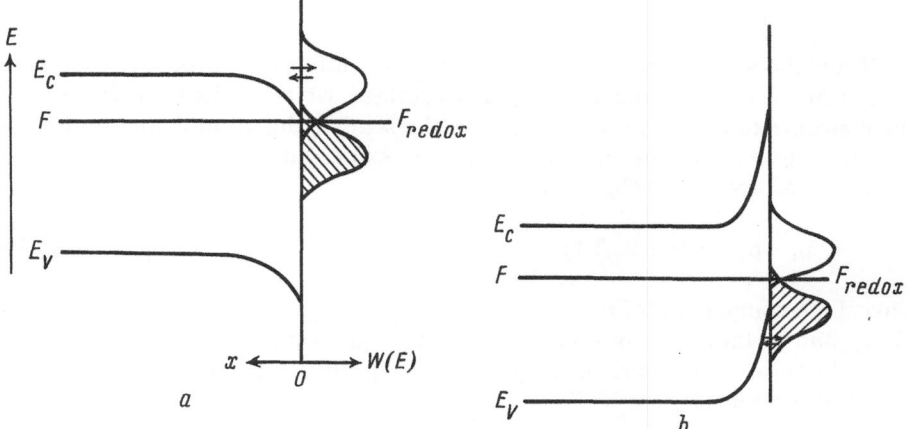

Fig. 18. Schematic of electronic and hole currents between semiconductor electrode and solution containing redox system with more negative (a) and more positive (b) reversible potential Electron transitions at the interface are shown by arrows

through the valence band we can write:

$$i_p^a = i_p^0 \left(\frac{p_s}{p_s^0} \ \frac{c_{red,s}}{c_{red,s}^0} \right) \exp\left(e\alpha_p\eta_H/kT\right) \tag{1.26}$$

Here, i^0 is the exchange current; p_s and $c_{red,s}$ are the concentrations of "reagents", namely of holes and Red-particles of the solution, respectively, at the interface; α is the transfer coefficient of the reaction (usually, $0 < \alpha < 1$). The superscripts a and o and subscript p stand respectively for "anode", "equilibrium", and "hole". The quantity $\eta_H = \Phi_H - \Phi_H^0$ is that part of total overvoltage[7], $\eta = \eta_H + \eta_{sc}$, which is accounted for by the Helmholtz layer; the other part of the overvoltage is accounted for by the space charge layer in the semiconductor, $\eta_{sc} = \Phi_{sc}^0 - \Phi_{sc}$.

Neglecting deviation in the concentration of the substance to be oxidized (Red) in solution from the equilibrium concentration and allowing for the inverse reaction, we obtain a complete equation for hole current:

$$i_p = i_p^0 \left\{ \frac{p_s}{p_s^0} \exp\left(e\alpha_p\eta_H/kT\right) - \exp\left[-e(1-\alpha_p)\eta_H/kT\right] \right\} \tag{1.27}$$

An analogous equation for electronic current passing through the conduction band is:

$$i_n = i_n^0 \left\{ \exp\left(e\alpha_n\eta_H/kT\right) - \frac{n_s}{n_s^0} \exp\left[-e(1-\alpha_n)\eta_H/kT\right] \right\} \tag{1.28}$$

Here, n is the concentration of electrons in the conduction band, and subscript n stands for "electronic".

Total current is the sum of electronic and hole currents:

$$i = i_p + i_n \tag{1.29}$$

Thus, the amount of flowing current significantly depends on the concentration of free carriers on the semiconductor surface, which, in its turn, depends on the concentration of the same carriers in the semiconductor bulk and on the potential drop Φ_{sc}. This dependence is usually expressed by the Boltzmann equation. For instance, for holes we have:

$$p_s = p_0 \exp\left(-e\Phi_{sc}/kT\right) \tag{1.30}$$

Thus, both components of total overvoltage appear (η_H directly, see Eqs. (1.27) and (1.28), and η_{sc} through p_s or n_s) in the complete kinetic equation, but with different coefficients (α or 1). The exact shape of the $i - \varphi$ curve depends on the relative values of $\Delta\Phi_H$ and $\Delta\Phi_{sc}$.

---

[7] Recall that overvoltage is equal to the deviation of electrode potential from the equilibrium potential: $\eta = \varphi - \varphi^0$.

From the foregoing it is evident, first, that the overvoltage of electrode reaction determines the kinetics of the reaction on a semiconductor electrode not to a less extent than on a metal electrode. And the exchange current of any reaction on a semiconductor electrode, as follows from the theory of the elementary act (see, for example, Sect. 4.4 in Ref [1]), is much less than on a metal electrode due to lower concentration of free carriers, i.e., the reaction is less reversible. All this forces one to consider with utmost care any possibility of lowering the overvoltage (using, say, different types of catalysts, mediators, etc.). In particular, the reaction will proceed easily if impurities (or defects) are introduced onto the semiconductor electrode surface, whose energy levels when located in the forbidden band of the semiconductor satisfactorily match with a corresponding distribution of levels in solution (in case this distribution badly matches with the allowed energy bands in the semiconductor).

Second, the dependence of current on potential is very sensitive to (as discussed in the previous section) the phenomenon of band-edge pinning or Fermi-level pinning on the semiconductor surface, because the relative values of $\Delta\Phi_{sc}$ and $\Delta\Phi_H$ are subject to this phenomenon.

Finally, the location of reversible potential (or, the same, F_{redox}) relative to the band edges in the semiconductor is of great significance for a particular mechanism of the reaction.

Note that, besides the briefly discussed approach which may be called "kinetic" and amounts to quantitative determination of the current of electrode reaction in one model or another (Eq. (1.27) taken as an example), there exists another, thermodynamic method for qualitative description of the same reaction. The latter method makes use of the concept of the electrochemical potential of a redox system. Though electrons transfer, of course, only between the allowed energy levels (filled or vacant) in the contacting phases (see Fig. 18), the kinetic behavior of the system can be characterized also by the electrochemical potential levels in the phases, F and F_{redox} (despite the fact that there are no electrons, for instance, at the Fermi level of the semiconductor). It is precisely in analogy with the equilibrium condition, $F = F_{redox}$ (or, the same, $\varphi = \varphi^0$) that the condition for a reaction to occur on the anode can be expressed as:

$$F < F_{redox}, \quad \text{or} \quad \varphi > \varphi^0 \tag{1.31a}$$

For a reaction to occur on the cathode it is necessary that:

$$F > F_{redox}, \quad \text{or} \quad \varphi < \varphi^0 \tag{1.31b}$$

If one of the conditions of Eqs. (1.31) is fulfilled, then at least a corresponding reaction is thermodynamically possible (though the aforementioned kinetic limitations may, of course, strongly retard or even completely rule out the possibility of this reaction).

Chapter 2
Processes Underlying the Action of Semiconductor Photoelectrochemical Cells

2.1 Fundamentals of Describing Photoelectrochemical Reactions on Semiconductor Electrodes

The processes occurring with the participation of non-equilibrium current carriers – electrons and holes – in an illuminated semiconductor underlie the action of photoelectrochemical solar cells. Among these processes are: photogeneration and separation of charges and their transfer from the semiconductor through the phase boundary (with the electrolyte, on the one hand, and with the ohmic contact metal, on the other hand). The most effective is light with wavelength less than the semiconductor fundamental absorption threshold; it causes generation of electron-hole pairs (see Sect. 1.2). The generation rate $g(x)$ (i.e., the number of non-equilibrium carriers appearing in a unit volume per unit time) is maximum near the semiconductor surface and decreases on going deep into the semiconductor as per the law (cf. Eq. (1.6)):

$$g(x) \sim \alpha J_0 \exp(-\alpha x) \tag{2.1}$$

where J_0 is the flux density of photons incident on the surface; α is the optical absorption coefficient. The $g(x)$ curve is shown in Fig. 19 (curve 1).

Let us isolate two regions in the semiconductor, namely, the space charge layer (of thickness L_{sc}) and the remaining uncharged bulk. In the former, the current carriers drift mainly in the electric field. In the latter, as there is no field, they are transported as per the diffusion mechanism. For example, in a n-type semiconductor with a band bending resulting in the appearance of a depletion layer (Fig. 19 shows precisely this case) the minority carriers (holes) in the space charge electric field are transferred to the semiconductor surface, and majority carriers (electrons), to the semiconductor bulk. This is how separation of charges takes place (this is the most important and most typical function of the semiconductor used for light-energy conversion).[1]

Assume that on the semiconductor surface ($x = 0$) the holes are effectively "trapped" due to the occurrence of a rapid electrochemical reaction on this sur-

---

[1] The completeness of such a separation depends on the strength of the field. In weak fields (when the thickness of depletion layer is large and the band bending is small), part of the majority carriers reach the semiconductor surface by diffusing against the field. This causes the photocurrent to decrease (see, for instance, Ref. [31]).

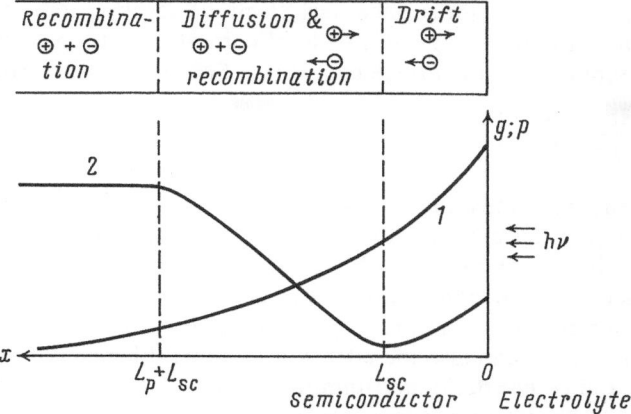

Fig. 19. Current carriers photogeneration rate (1) and minority carrier concentration distributions (2) in a semiconductor with the surface illuminated. At top are marked the main processes involving participation of current carriers

face and/or due to surface recombination. In this case, the surface acts as a "drain" for holes and their actual surface concentration $p(x = 0) = p_s$ is less than the equilibrium surface concentration (see Eq. (1.30)), though it is more than at the inner boundary of the space charge region ($x = L_{sc}$) because of the action of the electric field.

At the inner boundary of the space charge region, which acts as a "drain" for holes in the uncharged bulk of the semiconductor, the concentration of holes $p(L_{sc})$ is less than deep in the semiconductor, $p(\infty) = p_0$. Here, the holes diffuse under the action of concentration gradient and recombine at the same time. The combined action of enumerated factors in the near surface region of the semiconductor yields the hole-concentration profile shown in Fig. 19. This distribution resembles the concentration distribution of an ion reacting on the electrode at a relatively low total concentration of the electrolyte in solution [32]. The depletion layer range is equivalent to the diffuse ionic layer (Gouy layer) and the diffusion length of holes L_p (i.e., the average distance travelled by a non-equilibrium hole before it recombines with an electron) plays the role of diffusion layer thickness. Recall that $L_p = \sqrt{D_p \tau_p}$, where D_p and τ_p are respectively the diffusion coefficient and the lifetime of holes.

In the first approximation, the drift time in which the non-equilibrium carriers pass through the space charge layer is much less than their lifetime. Therefore, recombination of holes in this region may be neglected. This implies that all holes generated within the depletion layer reach the electrode surface and may contribute to the photocurrent. But the recombination of holes in the semiconductor bulk cannot be neglected; its intensity depends on the diffusion length L_p. The holes photogenerated at a distance more than $L_{sc} + L_p$ from the surface can no longer reach the surface: they recombine earlier without participating in the transfer of current through the semiconductor/electrolyte interface.

For the simplest case (when recombination does not take place in the space charge region and on the semiconductor surface; fast electrode reaction traps all holes that reach the surface) the photocurrent, i.e., total current of holes passing through the illuminated semiconductor surface equals (see Sect. 6.1 of Ref. [1]):

$$i_{ph} = eJ_0 \left[1 - \frac{\exp(-\alpha L_{sc})}{1 + \alpha L_p} \right] \tag{2.2}$$

Measuring the photocurrent experimentally we can determine any of the values α, L_{sc} or L_p if the other two quantities are measured by an independent method.

In more complex theories, other effects are taken into consideration, such as recombination in the space charge region and on the electrode surface, the afore-mentioned diffusion of majority carriers in the space charge layer, finite reaction-rate at the semiconductor surface, and also the effects of ultimate thickness of the semiconductor electrode and of nonstationary illumination (see Sect. 6.1 in Ref. [1] and Refs. [31, 33–39]). The outcome of these calculations can be formulated as follows: each of the enumerated effects can be separately allowed for and an expression for the photocurrent can be obtained in the analytical or numerical form. Simultaneous consideration of several of these effects yields very awkward formulas. And *a priori* it is not always known what "weight" these effects will have in the total photocurrent. Equation (2.2) is best suited for practical purposes.

As in the case of "dark" electrode reactions (see above), besides the briefly expounded kinetic approach to the quantitative description of photoelectrochemical reactions, which amounts to computing in one or another model the semiconductor electrode photocurrent, use is often made of the quasi-thermodynamic approach for qualitative description of photoelectrochemical reactions on semiconductors. By the latter approach (Sect. 6.3 in Ref. [1]), in an unilluminated semiconductor at equilibrium the electrons are characterized by an electrochemical potential (Fermi level). Under non-equilibrium conditions, in particular, in the illuminated semiconductor, the generated free carriers, electrons, and holes are no more in equilibrium with each other, and, therefore, no electrochemical potential can yet be assigned to the semiconductor on the whole. Nevertheless, under certain conditions both electron and hole ensembles, each taken separately, can be assumed to be at equilibrium. This means that by exchanging energy, the initial ("dark") carriers of the band strike a balance with the photogenerated carriers. Since an equilibrium between the bands is still absent, such a state of the semiconductor is called the quasi-equilibrium state. Then for every type of current carriers – holes in the valence band and electrons in the conduction band – a separate "electrochemical potential" level (the so-called quasi-Fermi level) can be introduced, which describes the properties of the ensemble in the same manner as the Fermi level describes those of the electron ensemble of the equilibrium semiconductor on the whole. The shift of the quasi-Fermi level, say, of holes F_p in the illuminated semiconductor relative to F of the same semiconductor at equilibrium (Fig. 20) depends on the concentration of the non-equilibrium holes (which, in its turn, is the function of light intensity, light absorption coefficient of the semiconductor, recombination velocity, and other factors):

$$F - F_p = kT \ln \frac{p_0 + \Delta p}{p_0} \tag{2.3}$$

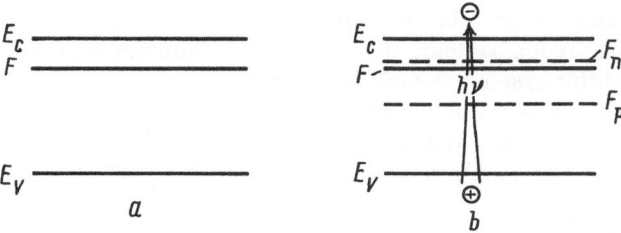

Fig. 20. Cleavage of semiconductor Fermi level into quasi-levels of electrons and holes (a) semiconductor under equilibrium (in the dark); (b) semiconductor in the quasi-equilibrium state (upon illumination)

where p_0 is the dark concentration of holes and Δp is the increase in concentration upon illumination. Depending on the relationship between p_0 and Δp (or between n_0 and Δn for electrons) the quasi-Fermi level shift may be large (for minority carriers whose dark concentration is often vanishingly small) or almost insignificant (for majority carriers whose concentration is scarcely affected by illumination). Thus, in a wide bandgap n-type semiconductor $F_n \simeq F$, while the quantity $F - F_p$ may account for a large fraction of the semiconductor bandgap width.[2]

The condition for a photoanodic reaction involving the participation of holes to occur is now, instead of Eqs. (1.31), written as:

$$F_p < F_{redox} \tag{2.4a}$$

and for the photocathodic reaction involving the participation of conduction band electrons, this condition takes the form:

$$F_n > F_{redox} \tag{2.4b}$$

Returning to the relationship between the electrochemical potential level and the electrode potential, it can be concluded that the semiconductor electrode in a quasi-equilibrium state is as if simultaneously at two different potentials, of which one controls the reaction occurring via the conduction band, and the other, the reaction proceeding via the valence band. This unique property of semiconductor electrodes has no analogy in the electrochemistry of metals.[3]

Using the concept of quasi-Fermi level, we shall qualitatively examine how the most important types of photoelectrochemical reactions proceed on semiconductors. To this end, we shall first turn to the case where a semiconductor and a solu-

---

[2] The quasi-Fermi levels are formed in that region of the semiconductor body where the concentration of current carriers is different from the equilibrium one. Upon illuminating the semiconductor surface this region has thickness of order $\alpha^{-1} + L_p$.

[3] The Fermi level F of a semiconductor electrode can be directly measured as its potential. And the location of the quasi-Fermi level of minority carriers can be judged only by indirect criteria (say, by the equilibrium potential of a redox reaction which upon illumination proceeds on this electrode with the participation of minority carriers).

tion in the dark are already in equilibrium. This happens when the solution contains a highly reversible redox system; here, the semiconductor assumes the equilibrium potential of this system: $\varphi = \varphi^0$, or $F = F_{redox}$.

Let us now analyze in detail the action of such a photoelectrochemical cell by considering an example of the system: n-type GaAs/alkaline solution of $Se^{2-} + Se_2^{2-}$/metal cathode. An equilibrium is established in the cell in the dark:

$$Se_2^{2-} + 2\,e^- \rightleftarrows 2\,Se^{2-}$$

Now both electrodes have attained the equilibrium potential of this redox system, so that the Fermi level of the metal (F_{met}) and GaAs (F) and the level $F_{redox} = F_{Se^{2-}/Se_2^{2-}}$ in the solution become equal (Fig. 21a). The equilibrium potential is taken more positive than the flat band potential of the semiconductor, $\varphi_{Se^{2-}/Se_2^{2-}}^0 > \varphi_{fb}$, so that a depletion layer is formed in the semiconductor, which ensures better separation of charges. In the illuminated semiconductor, the photogenerated electrons and holes in the electric field of the depletion layer move in mutually opposite directions: in the case discussed here, the holes move towards the interface, and the electrons, into the semiconductor bulk (Fig. 21b). The electric field resulting due to this separation partially compensates the initial field, and the band bending $e|\Phi_{sc}|$ decreases upon illumination, i.e., the bands unbend (flatten). Unbending of bands, in its turn, causes other energy levels in the system to change their location. In fact, the location of the Fermi level F relative to the band edges in the semiconductor bulk is strictly given (see Eq. (1.4)). Therefore, upon unbending in the light, the bands "pull" the Fermi level; the latter shifts relative to its location in the electrode in the dark, as shown in Fig. 21b. This shift ΔF, as already mentioned, may be directly measured in the cell by using a reference electrode, namely $\Delta F = e|\varphi_{ph}|$, where φ_{ph} is the photopotential, i.e., the dif-

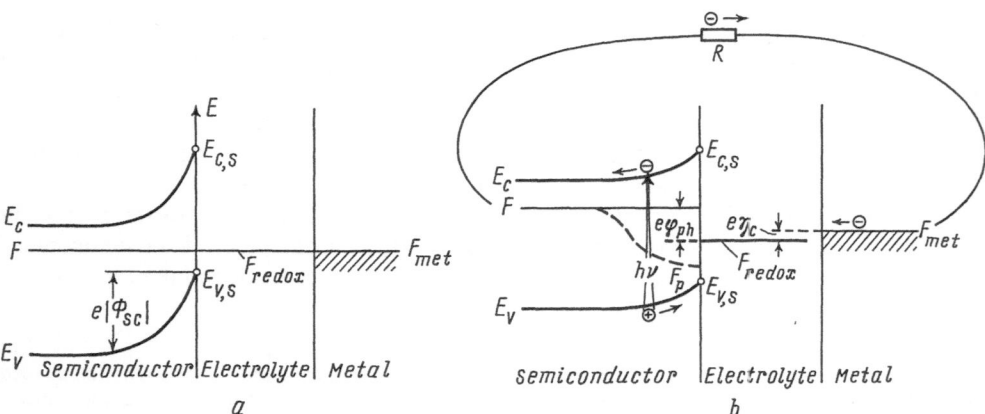

Fig. 21. Energy diagram of photoelectrochemical cell with a n-type semiconductor photoanode, metal cathode and solution containing a reversible redox system
(a) in the dark, with open circuit; (b) upon illumination, short-circuited via load resistance
η_c – overvoltage of reaction at metal counter electrode

ference in electrode potentials in the light and in the dark. (Here, for simplicity, the band-edge pinning at the surface case has been considered, that is, the potential drop in the Helmholtz layer does not vary upon illumination, and the entire potential change is concentrated in the space charge layer of the semiconductor.) The photopotential is related to light intensity J_0 (at moderate J_0) as:

$$|\varphi_{ph}| = \gamma \frac{kT}{e} \ln\left(\frac{i_{ph}}{i_s} + 1\right) \sim \gamma \frac{kT}{e} \ln J_0 \qquad (2.5)$$

Here, i_{ph} is the anodic photocurrent; i_s is the dark locking current which in the order of magnitude is equal to the exchange current of majority carriers i_n^0 (see Sect. 1.5 of this book); and γ is a coefficient, often known as "quality factor", which allows for the deviation in the behavior of photoelectrode from the ideal case where $\gamma = 1$.

The quasi-Fermi levels of minority and majority carriers, F_p and F_n, are formed upon photogeneration of electron-hole pairs in the near-surface layer of the illuminated semiconductor. Since $F_p < F_{redox}$ and $F_n \simeq F > F_{redox}$, both forward and reverse reactions in the redox system must accelerate in the light. When the cell circuit is closed with the external load R, the electrons transfer from the semiconductor to the metal cathode of the cell where Se_2^{2-} ions are reduced to Se^{2-}; the holes go from the semiconductor photoanode to the solution, as a result of which the Se^{2-} ions are oxidized to Se_2^{2-}. The composition of the solution does not change on the whole while the current flows via the external circuit of the cell, i.e., conversion of light energy into electrical energy takes place. Under the band-edge pinning conditions, the maximum value of $|\varphi_{ph}|$ (i.e., the open-circuit photopotential) equals in the absolute value, as is seen from Fig. 21b, the potential drop in the space charge layer $|\Phi_{sc}|$, i.e., the difference:

$$|\varphi_{ph}^{o.c}| = |\varphi^0 - \varphi_{fb}| \qquad (2.6)$$

To raise the energy conversion efficiency of a photoelectrochemical cell, the semiconductor and the redox system in solution should be so selected that the difference $|\varphi^0 - \varphi_{fb}|$ is as much as possible.

Let us now consider a case where the reaction on the cathode is not the reverse reaction with respect to that occurring on the photoanode, i.e., two different reactions take place on the cell electrodes upon illumination. This is possible, in principle, if none of the redox systems present in the solution is well reversible on the semiconductor electrode. Then, equilibrium is not established in the cell in the dark. The semiconductor electrode behaves as an ideally polarizable one and its stationary potential is determined, for example, by adsorption of traces of oxygen. This is how, say, the semiconductor oxides behave in solutions of electrochemically inactive electrolytes. The important condition is that a depletion layer necessary for separating the photogenerated carriers should appear in the semiconductor at a stationary potential.

Figure 22 shows the energy diagram of a photoelectrochemical cell with such a semiconductor photoanode, metal cathode, and aqueous electrolyte solution. The

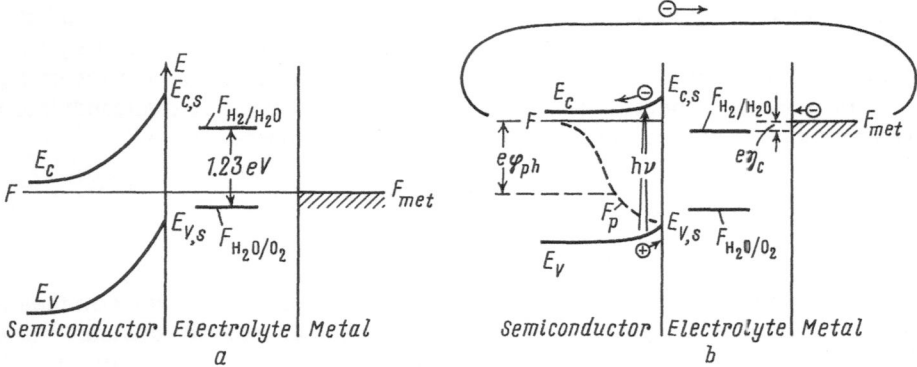

Fig. 22. Energy diagram of photoelectrochemical cell with a n-type semiconductor photoanode, metal cathode, and solution containing no reversible redox system
In the solution are shown the electrochemical potential levels of water oxidation and reduction reactions
(a) in the dark, with open circuit; (b) upon illumination, short-circuited

electrochemical potential levels for the oxidation and reduction reactions of water:

$$2\,H_2O + 2\,e^- = H_2\ + 2\,OH^- \quad F_{H_2/H_2O}$$
$$H_2O + 2\,h^+ = \tfrac{1}{2}O_2 + 2\,H^+ \quad\ \ F_{H_2O/O_2}$$

$$(2.7)$$

are shown in the figure. (Some other reactions that may possibly occur in this system are considered in the next section.) With respect to these levels, the Fermi level of short-circuited semiconductor anode and metal cathode in the dark takes an arbitrary position. Electrons and holes are generated upon illuminating the semiconductor, which, as discussed earlier, separate in the depletion layer electric field, the bands unbend, and the Fermi level F shifts upwards and together with it moves also the quasi-level of electrons F_n (as is shown in Fig. 22 b). The quasi-level of holes F_p lowers down, on the contrary: the higher the photogeneration rate of holes, the lower the quasi-level goes.

At a certain illumination intensity, the quasi-Fermi level of holes F_p on the semiconductor surface may attain the water oxidation-reaction level F_{H_2O/O_2} and the quasi-level of electrons F_n achieves, together with the Fermi level F, the water reduction level F_{H_2/H_2O}; the latter reaction moves to the metal cathode when the external circuit closes. These two reactions start proceeding simultaneously in the cell. Thus, the non-equilibrium electrons and holes generated in the semiconductor upon illumination are spent to stimulate the corresponding partial reactions, Eq. (2.7), which jointly constitute the process of decomposing water into hydrogen and oxygen:

$$H_2O\ \rightarrow\ H_2 + \tfrac{1}{2}O_2$$

$$(2.8)$$

Conversion of light energy into chemical energy of the photoelectrolysis products thus takes place.

Instead of decomposition of water in the cell, another process, say, photocorrosion of the semiconductor electrode, may also proceed (see the following section). With the conduction band bottom $E_{C,s}$ located above the F_{H_2/H_2O} level, as shown in Fig. 22, the energy of electrons in the conduction band (or in the metal cathode) suffices to reduce water; hence, upon illumination, this reaction proceeds spontaneously. More precisely, for a n-type semiconductor, this reaction proceeds spontaneously if $\varphi_{fb} < \varphi^0_{H_2/H_2O}$; for a p-type semiconductor photoelectrode, this happens if $\varphi_{fb} > \varphi^0_{H_2O/O_2}$. If these conditions are not fulfilled, then the energy of majority carriers is insufficient for a corresponding partial reaction to occur; photodecomposition of water is possible only if, besides illuminating the photoelectrode, a voltage is applied to the cell from an external source. (In this case, all energy levels in the semiconductor will shift relative to the solution.)

Comparing the considered two types of photoelectrochemical processes underlying the cells that are employed to convert light energy into electrical energy and chemical energy, we must note their following common features:
a) the electric field of the depletion layer in a semiconductor is used to separate the photogenerated charges;
b) the electrochemical reaction that proceeds with the participation of majority carriers moves to the metal counter-electrode, which ensures (1) spatial separation of the photoelectrolysis products and, (2) some decrease in energy losses due to overvoltage even if just for one of the partial reactions (because a metal electrode is generally superior to a semiconductor electrode in electrocatalytic activity).

Now we shall make two concluding remarks on the aforementioned quasi-thermodynamic approach to the description of photoelectrochemical reactions. First, this approach is based on the assumption that equilibrium has been reached within the electron and the hole ensemble. In other words, the photoexcited current carriers have had time to transmit their kinetic energy (approximately equal to the difference $h\nu - E_g$) to other carriers by interacting with them, i.e., the current carriers have thermalized. The conditions under which this can happen are discussed in detail in Sect. 6.4 of Ref. [1]. In practice, a situation is apparently met with [40] when this process does not manage to occur and the carriers come to the electrode surface while "hot". Excess energy permits them to transfer to higher energy levels in the solution than do thermalized carriers (having energy close to the edges of the allowed bands on the surface: $E_{C,s}$ and $E_{V,s}$). As a result of this, electrode reactions which cannot occur under quasi-equilibrium conditions (in particular, the reactions whose energy levels in solution lie above $E_{C,s}$ or below $E_{V,s}$) may proceed involving the participation of "hot" electrons or holes.

The other remark concerns the earlier made assumptions of band-edge pinning on a semiconductor surface in the dark as well as upon illumination. In Sect. 1.4 we have seen that on varying the electrode potential the band edges remain pinned on the surface (i.e., the Helmholtz potential drop remains unchanged) only in a limited range of potentials. Beyond the limits of this range the potential drop in the Helmholtz layer starts changing and the bands unpin.

The same happens when the semiconductor electrode potential is varied upon

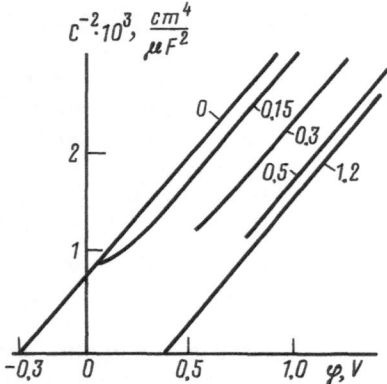

Fig. 23. Shift of Mott-Schottky plct for n-type WSe$_2$ electrode in 2 M HCl upon illumination [41]

Illumination intensity (in relative units) is indicated on curves

illumination (i.e., when a photopotential appears) and for the same reasons (high density of surface states, etc.). It does not matter how the electrode potential is varied (i.e., by using an external voltage source or by illumination) for the potential to redistribute between the space charge layer and the Helmholtz layer.

By way of illustration, unpinning of bands upon illumination is depicted in Fig. 23. Therein the Mott-Schottky plot is shown for a n-type WSe$_2$ electrode both in the dark and upon illumination. If flattening of bands (i.e., change in Φ_{sc}) would have been the only outcome of illumination and the potential drop Φ_H had remained constant, then on increasing the light intensity J$_0$ the open circuit electrode potential would have changed by $\varphi_{ph}(J_0) = |\Delta\Phi_{sc}|$ and the point characterizing the semiconductor electrode would have shifted along the straight line ($C^{-2} - \varphi$ plot) measured for the electrode in the dark by varying its potential with the aid of an external source. In the considered experiment, this happens only in the range of potentials from -0.3 to $+0.1$ V. At more positive potentials the $C^{-2} - \varphi$ plot taken for the illuminated semiconductor electrode shifts parallel to itself with respect to the line measured for the electrode in the dark. The stronger the illumination, the greater the shift. In other words, the flat band potential (obtained by extrapolating the $C^{-2} - \varphi$ straight line to $C^{-2} \rightarrow 0$) upon strong illumination is found to be more positive by 0.7 V than in the dark. This is probably due to a change in the charge of the surface states that causes the potential drop in the Helmholtz layer to vary (see Sect. 1.4). Other semiconductor electrodes, say GaAs [42], CdTe [43], and GaP [44], behave in the same manner as WSe$_2$.

Unpinning of band edges, whatever may be the cause, whether observed upon varying the potential of the redox system in solution or upon illumination limits the maximum possible value of photopotential, i.e., the photopotential at complete flattening of bands. This, as shown above, equals $|\varphi_{ph}| = |\Phi_{sc}| = |\varphi^0 - \varphi_{fb}|$ when the Helmholtz potential drop remains constant. But if the potential drop in the Helmholtz layer Φ_H increases, then the potential drop in the space charge layer Φ_{sc} decreases at constant electrode potential. Correspondingly, $|\varphi_{ph}|$ also decreases. This reduces the efficiency of photoelectrochemical cells for converting light energy into electrical energy.

2.2 Photocorrosion and Protection of Semiconductor Electrodes

A disadvantage of photoelectrochemical cells with semiconductor electrodes is the corrosion of these electrodes. Corrosion is particularly intense on illuminated electrodes. (This phenomenon is known as photocorrosion.) Many semiconductor compounds decompose during anodic as well as cathodic polarization. The cathodic decomposition reaction usually proceeds with the participation of conduction band electrons, and the anodic decomposition reaction, with the involvement of valence band holes.[4] For instance, during strong cathodic polarization of some electrodes (e.g., ZnO, CdS) in aqueous solutions the electrodes decompose with the deposition of metal on the electrode surface; during anodic polarization the electrode material decomposition is accompanied by the formation of a non-conductive oxide film on its surface (e.g., on Si) or by the transfer of metal ions to the solution (e.g., in the case of GaAs or CdS). These processes significantly limit the service life of photoelectrochemical cells.

The predisposition-to-corrosion criteria can be conveniently expressed in the qualitative form, using the quasi-thermodynamic approach (Sects. 7.1 and 7.2 in Ref. [1]) and considering corrosion as an example of redox reactions (see Sect. 2.1 of this book). Thus, the electrochemical potentials $F_{dec,n}$ and $F_{dec,p}$ related to the corresponding equilibrium potentials by Eq. (1.13) may be ascribed to the cathodic and anodic reactions proceeding with the involvement of electrons and holes, respectively. The values of $F_{dec,n}$ and $F_{dec,p}$ computed from thermodynamic data are shown in Fig. 24 for several important semiconductor electrode materials. The thermodynamic condition for possible occurrence of corrosion reactions by analogy to Eq. (1.31) is expressed as:

for anodic decomposition $\quad F < F_{dec,p}, \quad$ or $\quad \varphi > \varphi^0_{dec,p}$ $\qquad$ (2.9a)

for cathodic decomposition $\quad F > F_{dec,n}, \quad$ or $\quad \varphi < \varphi^0_{dec,n}$ $\qquad$ (2.9b)

Thus, if during polarization of a semiconductor electrode the Fermi level of the electrode attains the electrochemical potential level $F_{dec,n}$ or $F_{dec,p}$, then the corrosive destruction of the electrode is to be expected. (Nevertheless, it must be remembered that in practice, by virtue of kinetic limitations, the reaction may be inhibited. Therefore, in many cases, just the kinetic but not the thermodynamic pecularities of an electrode reaction actually determine the corrosion and photocorrosion behavior of semiconductor electrodes.)

In the simplest case (band-edge pinning at the surface) the range of possible variations of the Fermi level F, and hence the range of achievable electrode potential is restricted by the semiconductor forbidden band. Therefore, if the electrochemical potential level $F_{dec,n}$ or $F_{dec,p}$ in the energy diagram faces the conduction or valence band, then this level cannot be attained, and the semiconductor is ther-

---

[4] The participation of holes in anode corrosion partial reactions, as in the process of anodic dissolution of semiconductors, amounts to that the localization of a hole on the interatomic bond weakens the bond and alleviates its rupture.

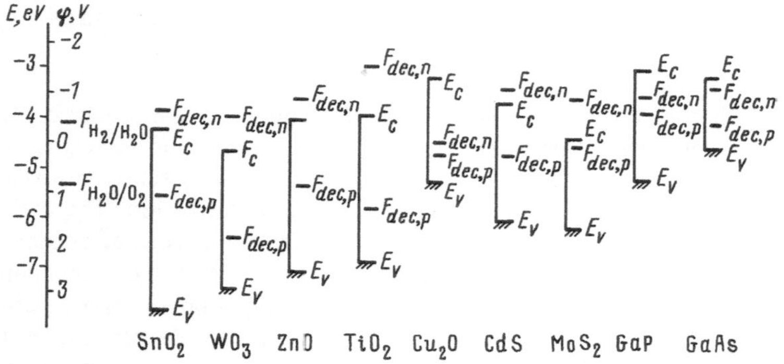

Fig. 24. Computed electrochemical potential levels for anodic and cathodic decomposition ($F_{dec,p}$; $F_{dec,n}$) reactions of semiconductor materials in comparison with conduction band bottom and valence band top energies (E_C; E_V) and electrochemical potential levels of water reduction and oxidation reactions (F_{H_2/H_2O}; F_{H_2O/O_2}) [45]

modynamically protected against corrosion. Thus, the semiconductor is absolutely protected from corrosion when the levels of both decomposition reactions lie beyond the forbidden band. As is seen from Fig. 24, more common are the cases when a semiconductor is resistant either to cathodic (e.g., SnO_2, WO_3, TiO_2, and MoS_2) or anodic decomposition. If both $F_{dec,n}$ and $F_{dec,p}$ are in the forbidden band (as in the case of Cu_2O, GaP, GaAs), then during anodic as well as cathodic polarization the semiconductor may decompose in principle.

Photocorrosion is considered in a like manner. Within the framework of the quasi-thermodynamic approach, the shift of the quasi-Fermi levels of electrons F_n and holes F_p in the illuminated semiconductor causes the decomposition reaction to accelerate. Therefore, Eq. (2.9) should be modified. For the semiconductor anodic photodecomposition reaction to proceed with the participation of holes, it is necessary that the condition:

$$F_p < F_{dec,p} \tag{2.10 a}$$

be fulfilled; the condition:

$$F_n > F_{dec,n} \tag{2.10 b}$$

should be complied with for the semiconductor cathodic decomposition reaction to proceed with the participation of conduction band electrons.

The described approach suggests at least, in principle, a way for avoiding corrosion (photocorrosion). Here, it must be taken into consideration that other reactions competitive with photodecomposition can also proceed. For example, as the quasi-levels F_n or F_p continue to shift with increasing illumination intensity, of all possible cathodic reactions, the reaction with a more positive equilibrium potential may start first, and, of all possible anodic reactions, the reaction with the most negative equilibrium potential may proceed first. Hence, the semiconductor corrosion can be suppressed by adding suitable oxidants or reductants to the solution. Thus, anodic photocorrosion can be avoided by using reductants which oxidize more readily than the semiconductor material, i.e., whose oxidation potential is more negative than the anodic decomposition potential: $\varphi^0 < \varphi^0_{dec,p}$, or $F_{redox} > F_{dec,p}$. The oxidants which are reduced more readily than the semiconductor material block out cathodic decomposition: $\varphi^0 > \varphi^0_{dec,n}$, or $F_{redox} < F_{dec,n}$. By oxidizing or reducing, such sacrificial system-protectors stabilize the semiconductor photoelectrodes, and, therefore, find application in photoelectrochemical cells for converting light energy into electrical energy. One such cell with a n-GaAs anode in the Se^{2-} solution is briefly considered in the preceding discussion; other cells will be considered in Chap. 6. The solvent may act as protector as well. Thus, as is seen from Fig. 24, SnO_2, WO_3, and TiO_2 are prone to anodic decomposition in principle; however, in practice, they are protected against this decomposition by the oxidation reaction of water, which proceeds more easily ($F_{H_2O/O_2} > F_{dec,p}$).

As in the consideration of photoelectrochemical kinetics as a whole (see Sect. 2.1), the photocorrosion processes are described by the kinetic as well as quasi-thermodynamic approach. Precisely, the kinetic equation is written by assuming that photocorrosion proceeds following a certain mechanism (i.e., by presuming a definite number of stages in partial reactions, the composition of intermediate products, etc.). Then corrosion depends on process conditions such as concentration of the oxidant which causes corrosion, concentration of the protector-reductant that stabilizes the semiconductor, rate constants of individual stages, light intensity, etc. (see, for instance, Ref. [46]).

2.3 Types of Photoelectrochemical Reactions and Classification of Photoelectrochemical Cells

Depending on whether one and the same reaction or two different reactions proceed on an illuminated semiconductor electrode and unilluminated metal counter-electrode, the composition of the solution on the whole remains unchanged or on the contrary, varies (cf. Sect. 2.1). Nozik [47] has proposed the following classification for photoelectrochemical cells (see Fig. 25). According to this classification, all cells are divided into (1) regenerative photoelectrochemical cells, also known as "liquid-junction solar cells" or "electrochemical photovoltaic cells", in which the Gibbs' free energy G of the electrolyte solution does not vary, and (2) photoelectrosynthetic cells in which the Gibbs' energy of the electrolyte solution changes when in operation.

In the former, of all redox systems (including solvent, solutes, and the semiconductor electrode material) which in principle exist in the cell, only one – the most reversible – system is effective. The components of this system participate in the electrode reactions; the cathode reaction products get oxidized at the anode and the anode reaction products reduce at the cathode. This is what happens in the photoelectrochemical cell with a GaAs electrode in the ($Se^{2-} + Se_2^{2-}$) solution;

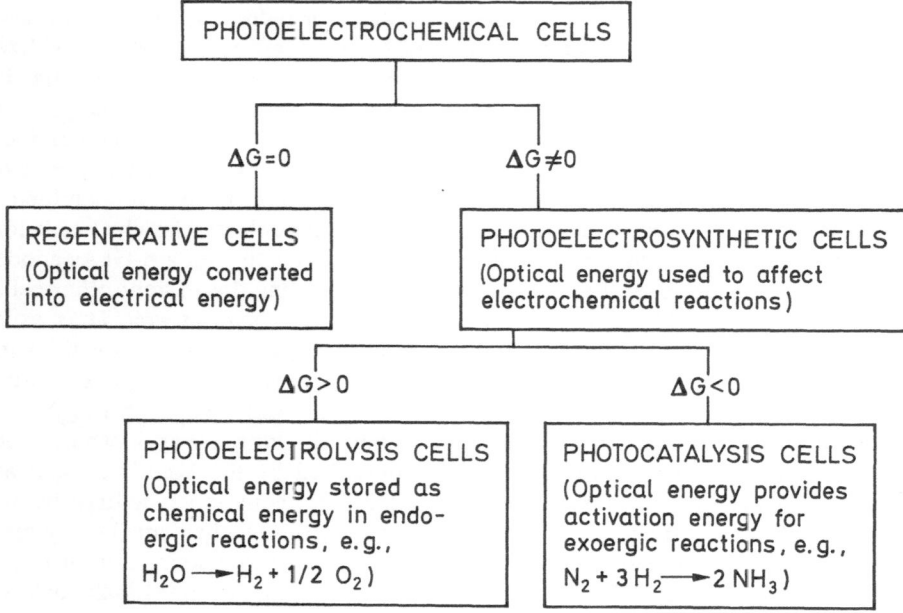

Fig. 25. Classification of photoelectrochemical cells (according to Ref. [47])

such a cell is considered in Sect. 2.1. As mentioned earlier, these photoelectro-chemical cells convert light energy into electrical energy.

In a photosynthetic cell, two different redox systems take part in complete reaction such that the anode reaction is not reverse with respect to the cathode reaction. Depending on the sign of ΔG, the cells can be for photoelectrolysis ($\Delta G > 0$) and photocatalysis ($\Delta G < 0$). In the former, the entire reaction proceeds in a "counter-thermodynamic" (uphill) way, namely by consuming light energy (but it does not proceed in the dark), therefore storage of energy takes place. Examples of this can be: photodecomposition of water into H_2 and O_2; photooxidation of halide ions; photoreduction of CO_2, and others. Conversely, storage of energy does not take place during photocatalysis and light energy is spent to overcome the activation barrier of the reaction which, in principle, could proceed in the dark, too (i.e., in a downhill way), but is inhibited in the kinetic respect (e.g., fixation of molecular nitrogen).[5]

During the different historic stages of developing semiconductor photoelectro-chemical cells for solar energy conversion, one or another type of cell received major attention of researchers. As a matter of fact, this history originated in the development of a cell for photoelectrolysis of water; this cell had a TiO_2 photo-anode [48]. The fate of this pioneering article is very curious. As seen from the ti-

---

[5] Note that in the photochemical literature the term "photocatalysis" is often used in a broad sense, denoting any photo(electro)chemical reaction.

tle "Electrochemical Evidence on the Mechanism of Primary Stages of Photosynthesis", the authors believed that the photoelectrochemical reactions (in particular, photoevolution of oxygen) occurring at the semiconductor (chlorophyll)/aqueous solution interface could be at the root of the photosynthesis process in green plants. Now it can be said with confidence that this viewpoint was incorrect, because there is no semiconductor phase as such in plants, and conversion of light energy with the participation of chlorophyll takes place at the molecular but not the phase level. Nevertheless, after 3–4 years this paper became of supreme concern to photoelectrochemists not in connection with the photosynthesis mechanism, but as an indication of the possibility of this new solar energy conversion method. The energy crisis of the mid-'70s added some psychological momentum to this. It may be said that this report, which acted as a "photocatalyst", called into being a new trend in semiconductor electrochemistry.

At first the interests of most researchers were concentrated upon photoelectrolysis of water, as a potential source of hydrogen for hydrogen energetics, and, in a more general manner, for the entire hydrogen economics of the future. By the end of the '70s, focus was on the study of regenerative cells and considerable success was achieved in improving the solar-energy conversion efficiency. These studies, in their turn, made it possible to tackle the photoelectrolysis problem in a novel manner, which is now again in the foreground.

2.4 Characteristics of a Photoelectrochemical Cell

The efficiency of a photoelectrochemical device for conversion of solar energy into electrical or chemical energy, according to the definition, equals[6]:

$$\eta = \frac{\text{Energy flux at outlet}}{\text{Energy flux at inlet}} \cdot 100\,\% \qquad (2.11)$$

Conditionally it can be represented as the product of several factors, each quantitatively taking account of a definite type of energy loss in the entire energy conversion process:

$$\eta = K_{thr} K_{st} Yf \cdot 100\,\% \qquad (2.12)$$

Let us now consider the physical meaning of these factors:

1. The quantity K_{thr} takes account of losses caused by the threshold (quantum) nature of light absorption in a semiconductor. As shown in Sect. 1.2., intrinsic absorption of light that gives rise to electron-hole pairs is possible only at such a quantum energy which, depending on the type of the interband transition, equals the forbidden bandwidth or slightly exceeds it: $h\nu \geq E_g$. For such an absorption, the conversion of energy of non-monochromatic light, as is sunlight (see Fig. 26),

---

[6] For the application of Eq. (2.11) in some particular cases, see Sect. 3.1 of this book.

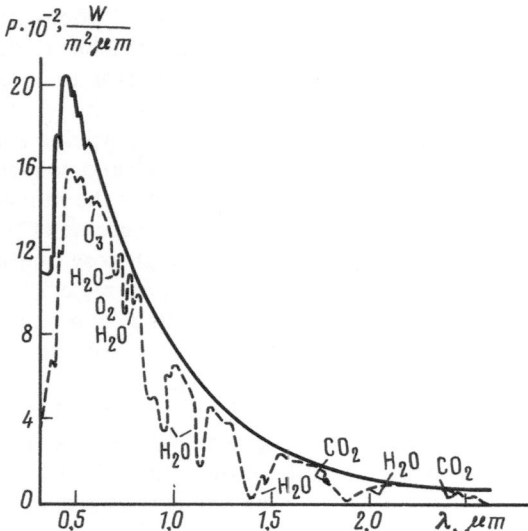

Fig. 26. Spectrum of sunlight Solid line – beyond the terrestrial atmosphere; dashed line – at the Earth's surface. The components of the Earth's atmosphere responsible for strong absorption of light are shown on the curve

entails unavoidable losses. In fact, quanta having energy less than E_g are simply not capable of generating electron-hole pairs. But quanta with energy exceeding E_g are not completely used: the excess energy dissipates, heating the semiconductor, but does not appreciably increase the number of current carriers.[7] Thus, for example, according to an estimate in Ref. [49], for silicon about 24 % of the energy of sunlight is lost due to the insufficient energy of photons and more than 32 % of the energy, being "excess quantum energy", changes into heat. The value of K_{thr} is determined by a particular spectrum of the radiation source and the chosen E_g:

$$K_{thr} = \frac{E_g \int\limits_{E_g}^{\infty} N(E)(1 - R)\, dE}{\int\limits_{0}^{\infty} EN(E)\, dE} \qquad (2.13)$$

Here, $N(E)$ is the amount of light quanta with energy $E = h\nu$ which fall on the semiconductor surface per unit time; $R(E)$ is the light reflection coefficient for the semiconductor surface. For a solar spectrum, the computed $(K_{thr} - E_g)$-type dependence is shown in Fig. 27. This dependence, as follows from the (considered in brief) physical nature of the process, takes the form of a curve having a maximum close to some optimal forbidden bandwidth $E_g^{opt} = 1.1$–1.5 eV (the exact value of E_g^{opt} slightly changes with variation of the spectral distribution, which, in its turn, depends on the "atmospheric mass", i.e., on the thickness of the atmospheric layer through which the sun rays pass).

---

[7] Here, the possibility of transition of *hot* carriers (see Sect. 2.1) through the interface is not allowed for (these transitions can increase K_{thr} somewhat).

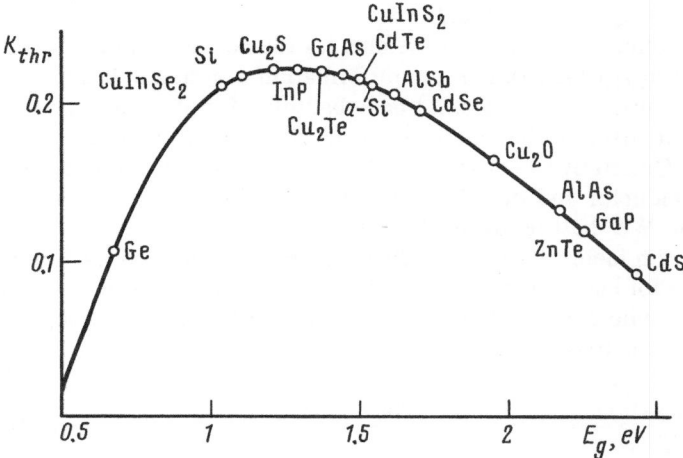

Fig. 27. K_{thr} versus forbidden bandwidth of semiconductor

The above-given interval of E_g^{opt} limits the range of semiconductor materials that can be used for making effective solar cells. More close to the "ideal" semi-conductors are: Si ($E_g = 1.11$ eV), InP ($E_g = 1.28$ eV), GaAs ($E_g = 1.43$ eV), and CdTe ($E_g = 1.50$ eV).

2. K_{st} characterizes the efficiency with which energy is stored, in other words, the efficiency of using the energy ($\simeq E_g$) of electron-hole pairs formed during photoexcitation, for doing useful work. In the photoelectrochemical reaction products at the outlet of the photosynthetic cell, energy equal to the variation of Gibbs' energy ΔG is accumulated. Therefore:

$$K_{st}^{ch} = \Delta G/E_g \qquad (2.14\,a)$$

Both in solid-state and photoelectrochemical solar cells for the conversion of light energy into electrical energy, it is the open-circuit photopotential $\varphi_{ph}^{o,c}$ which serves as a measure of useful work that can be done at the expense of the energy of excited current carriers. Then:

$$K_{st}^{el} = e\varphi_{ph}^{o,c}/E_g \qquad (2.14\,b)$$

Recall that the accessible ΔG is determined by the electrochemical potential of non-equilibrium current carriers in the illuminated semiconductor (see Sect. 2.1).

3. The quantum yield Y, according to definition, is the ratio of the number of electrons transferred through the external circuit of the cell (or consumed in the chemical reaction of forming new substances), to the number of light quanta incident on the photoconverter surface:

$$Y = i_{ph}/eJ_0 \qquad (2.15)$$

In the regenerative cells, i_{ph} is taken equal to $i_{sh.c}$ – the short-circuit photocurrent of the cell. In case of microheterogeneous systems for the conversion of solar energy into chemical energy (Chap. 5), the rate of formation of a new substance (in electrical units) in the primary stage should be substituted for photocurrent.

The quantum yield virtually characterizes the efficiency of separation of photogenerated charges. Qualitative knowledge of the dependence of Y on the properties of the semiconductor and of radiation can be obtained by analyzing Eq. (2.2) for photocurrent. We shall return to this later on.

4. Finally, the fourth factor, f – the so-called fill factor for the current-voltage characteristic – allows for energy losses caused by an ohmic voltage drop and also by overvoltages in the photoelectrochemical cell when current flows through it. For regenerative cells it equals:

$$f = \frac{(i_{ph} \cdot \varphi_{ph})_{MPP}}{i_{sh.c} \cdot \varphi_{ph}^{o.c}}$$ (2.16)

Here, the numerator shows the maximum electric power delivered by the cell (i.e., the maximum product of photocurrent and voltage, which can be obtained from the current-voltage characteristic) and the denominator, the product of the short-circuit current and the open-circuit voltage. The fill factor, f, depends on the shape of the current-voltage characteristic: the smaller the ohmic and other voltage losses, the more "rectangular" is the shape and the higher is f. In good quality solar cells, f reaches 0.70–0.75. Conversely, for large inner resistance of the cell, the shape of the characteristic is almost linear and f is small (0.30–0.25). The qualitative shape of the current-voltage characteristics for these two cases is shown in Fig. 28.

Let us now consider in brief the effect of semiconductor properties and the electromagnetic radiation spectrum on the enumerated characteristics of photoelectrochemical cells (see also Ref. [49]). The relationship between K_{thr} and the forbidden bandwidth is discussed earlier. The quantum yield of photocurrent, as follows from Eq. (2.2), is determined by the relationship between the light absorption coefficient, α, depletion layer thickness, L_{sc}, and diffusion length of minority carriers, L_p. The first of the enumerated quantities depends on the type of optical transitions in the semiconductor. L_{sc} depends on the concentration of majority carriers (see Eqs. (1.16) and (1.17)), which is controlled by introducing donor or acceptor dopants into the semiconductor. L_p depends on how perfect is the crystalline structure of the material and on the concentration of fortuitous impurities which act as recombination centers.

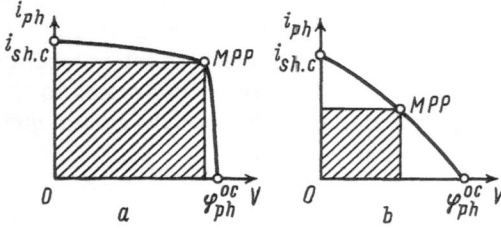

Fig. 28. Schematic representation of photocurrent-voltage curve for photoelectrochemical cell with large (a) and small (b) fill factor
Hatched areas are proportional to maximum output electric power.
MPP – the maximum power point on the characteristic

The optimal thickness of the depletion layer ranges between 10^{-6} and 10^{-5} cm; a concentration of majority carriers from 10^{15} to 10^{18} cm^{-3} corresponds to this range. At smaller L_{sc} the majority carriers can tunnel through the depletion layer to the solution instead of drifting in the depletion layer field towards the semiconductor bulk. At large L_{sc} the electric field strength in the depletion layer decreases such that the transit time of minority carriers becomes more than their lifetime and thus part of photogenerated carriers recombine within the depletion layer. Besides, the aforementioned undesirable diffusion of majority carriers against the electric field and their transfer to solution becomes quite probable.

When $\alpha^{-1} \ll L_{sc}$ the entire light is absorbed in the space charge region and photocurrent attains its maximum possible value eJ_0. Only the semiconductor with direct transitions, having $\alpha = 10^5$ cm^{-1} and more (see Fig. 4), satisfy this condition. In more commonly used semiconductors with indirect transitions, α usually does not exceed 10^4 cm^{-1} (and gradually increases with hv) such that $\alpha^{-1} \gg L_{sc}$ (cf. Fig. 19). Here, light is partly absorbed beyond the space charge region. This must be compensated by increasing the diffusion length; in fact, as follows from Eq. (2.2), in the examined case the photocurrent is proportional to $\alpha(L_{sc} + L_p)$. The increase in L_p calls for the use of an extra-pure material having perfect crystalline structure. This limits the possibility of using non-monocrystalline semiconductors which are not subjected to special purification and are therefore inexpensive.

The restrictions imposed on specific resistance of the electrode material (which depends on the concentration of the dopant) and also on the dimensions and shape of electrode follow, first, from the earlier considered condition for L_{sc} and also from the requirement that the ohmic potential drop in the cell should be small. As a guide, we list typical values of short-circuit photocurrent, $i_{sh.c}$: 10-30 mA/cm^2, and of open-circuit photovoltage, $\varphi_{ph}^{o.c}$: 0.3-0.7 V in the cells exposed to direct sunlight (i.e., $\simeq 100$ mW/cm^2). To avoid an appreciable decrease in f, the ohmic potential drop $i_{sh.c}R$ should be by an order of magnitude less than $\varphi_{ph}^{o.c}$. From this it follows that the total series resistance of the cell R should not appreciably exceed 1 ohm. This can be achieved by using semiconductor materials having resistivity of the order of 0.01-10 ohm · cm (of course, in the absence of significant resistance of current leads and overvoltage of the electrochemical reaction).

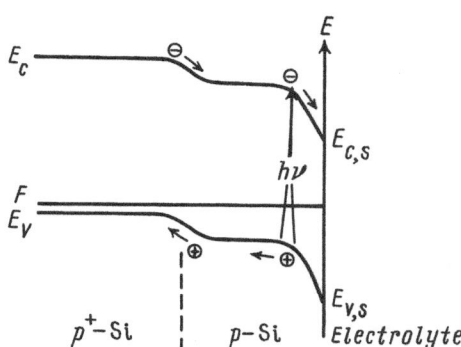

Fig. 29. Energy diagram of photocathode having p$^+$-p structure. Absorption of light and photogeneration of carriers take place in the less strongly doped near-surface p-layer. Strongly doped p$^+$-wafer serves as substrate

With the aim of reducing the series resistance, multilayer (e.g., epitaxial) semiconductor structures - the so-called $p^+ - p$ and $n^+ - n$ junctions - may be used as electrodes in which the outer layer, few microns thick, having relatively high resistivity (1-10 ohm $\cdot$ cm) ensures an optimal ratio between L_{sc} and α^{-1}, and the main part (substrate) which does not directly participate in light absorption and charge separation is made of a low-resistance material (hundredths and thousandths of ohm $\cdot$ cm). The band diagram of a like photocathode is shown in Fig. 29.

Chapter 3
Solar Energy Conversion into Chemical Energy.
Cells for Photoelectrolysis of Water

3.1 Basic Problems

Two conjugate electrochemical reactions which in the aggregate give rise to energy-rich products and are carried out at the expense of the energy of the electron-hole pair photogenerated in the semiconductor, form the basis of cells for the conversion of light energy into chemical energy. Historically, the first process of this kind (and the most important until the present time) is the photodecomposition of water into hydrogen and oxygen. Taking this as an example, we shall try to understand first the main characteristics and problems of photoelectrolysis. Other photoelectrochemical processes are considered in subsequent sections.

From the information listed in Sect. 2.1 on the operation of a photoelectrochemical cell with a semiconductor photoelectrode and a metal counter electrode, it is evident that the following conditions must hold for the spontaneous photoelectrolysis of water:

1. The energy of light quanta should exceed the forbidden bandwidth: $h\nu > E_g$.

2. The forbidden bandwidth should exceed the change of the Gibbs' free energy of reaction in the cell: $E_g > \Delta G$ (for decomposition of water $\Delta G = F_{H_2/H_2O} - F_{H_2O/O_2} = 1.23$ eV).

3. The flat band potential of n-type semiconductors (photoanode) should be more negative than the reversible potential of the hydrogen electrode reaction: $\varphi_{fb,n} < \varphi^0_{H_2/H_2O}$; when a p-type semiconductor (photocathode) is used, its flat band potential should be more positive than the reversible potential of the oxygen electrode reaction: $\varphi_{fb,p} > \varphi^0_{H_2O/O_2}$. Otherwise the energy of majority carriers will be insufficient for the partial reaction to proceed on the metal electrode of the cell.

4. The valence band edge on the n-type photoanode surface should be below the electrochemical potential level of the water oxidation reaction: $E_{V,s} < F_{H_2O/O_2}$ (for a p-type photocathode, $E_{C,s} > F_{H_2/H_2O}$). This is necessary so that the quasi-Fermi level of minority carriers (lying always within the forbidden band) could attain the electrochemical potential level of the corresponding partial reaction occurring on the photoelectrode.

5. Photocorrosion of the semiconductor photoanode can be prevented (see Sect. 2.2) if $\varphi^0_{H_2O/O_2} < \varphi^0_{dec,p}$ (for the photocathode, $\varphi^0_{H_2/H_2O} > \varphi^0_{dec,n}$).

The photoelectrolysis efficiency, according to Eq. (2.11), is expressed as:

$$\eta = \frac{\Delta G \cdot i_{ph}/e}{P_1} \cdot 100 \%$$ (3.1)

where P_1 is the incident light-flux power density.

It is not possible to make complete use of the energy of the formed electron-hole pair ($\simeq E_g$) because unavoidable losses of energy occur in the course of separation of charges and their transfer across the interphase boundaries. As an illustration, Fig. 30 shows the energy diagram of a cell with a n-type photoanode, whose characteristics are specially optimized to reduce these losses (in reality such a photoanode is yet not known), and a metal cathode (cf. Fig. 22). As is seen from this figure, the change of the Gibbs' energy equals only part of the forbidden bandwidth, namely:

$$E_g = \Delta G + e\,|\Phi_{sc}| + (E_C - F) + e\eta_c + (F_{H_2O/O_2} - E_{V,s})$$ (3.2)

Let us now enumerate the sources of energy losses:
1. Even under illumination, for effective separation of charges some finite potential drop must be retained in the depletion layer, $|\Phi_{sc}| > 0$. Usually, it suffices to have $|\Phi_{sc}| = 0.2-0.4\,\text{V}$.
2. The difference between the band edge of majority carriers and the Fermi level

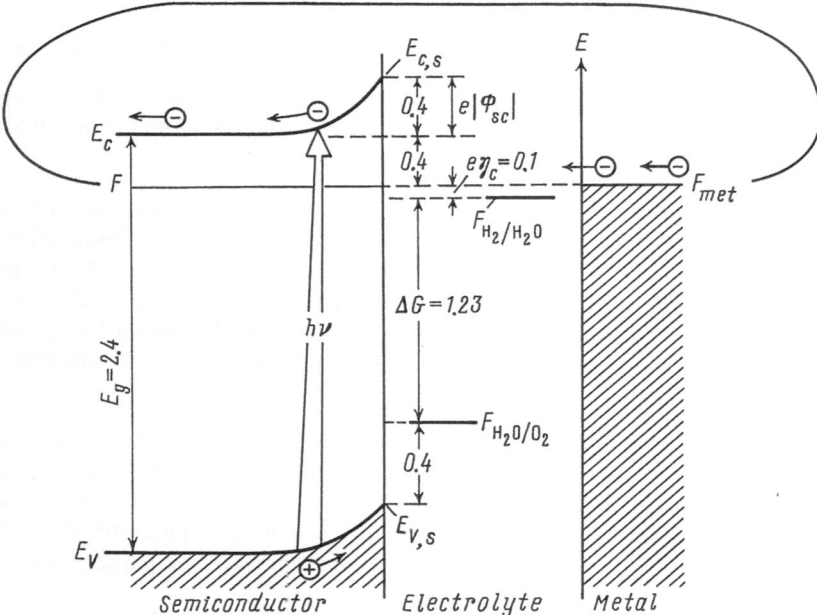

Fig. 30. Energy diagram of cell for photoelectrolysis of water with a n-type "optimized" semiconductor photoanode and metal cathode [50]. The state of short-circuited cell upon illumination is shown. The differences in energy levels (in eV) are listed in the figure

of the semiconductor, $E_C - F$, depends on the concentration of majority carriers (see Eq. (1.4)) which, according to the views expressed in Sect. 2.4, cannot be made very high. Generally, $E_C - F = 0.2\text{-}0.4\,\text{eV}$.

3. The overvoltage of a reaction on a metal cathode, η_c, depends on the electrocatalytic activity of the cathode. Usually one succeeds in lowering this overvoltage down to only 0.1 V.

4. The difference $(F_{H_2O/O_2} - E_{V,s})$ represents, first of all, the overvoltage of the reaction on the semiconductor anode, η_a, which, because of not very high electrocatalytic activity of semiconductors, can hardly be made less than 0.3-0.4 V. This difference depends, besides η_a, also on the mutual location of F_{H_2O/O_2} and $E_{V,s}$ determined by the chemical interaction between the semiconductor material and solvent and other factors which cannot be always influenced as desired.

The sum of all losses approximates to 1 eV. Whence, the forbidden bandwidth should exceed ΔG by this value, i.e., it should amount to about 2.2 eV.

Now we shall again examine the problem of optimal forbidden-bandwidth of a photoelectrode. In Sect. 2.3, in finding the dependence of K_{thr} on E_g (Fig. 27), we in fact assumed K_{thr} (Eq. (2.13)) and K_{st} (Eq. (2.14)) to be independent quantities. This assumption holds only for conversion of solar energy into electrical energy, because the latter can be transformed into useful work, in principle, at any potential. But photoelectrolysis occurs only at a quanta energy higher than some threshold equal to the sum of the Gibbs' energy change in a useful reaction and the aforementioned unavoidable losses. That is why the requirements to E_g become more stringent. Figure 31 shows the dependence of the upper limit of the theoretical efficiency of photodecomposition of water on the threshold wavelength λ_g conforming to the semiconductor forbidden bandwidth.[1] This dependence has been computed in Ref. [51] for an arbitrarily taken value of energy losses (cf. Fig. 27). For the "minimum" energy loss of 0.38 eV, the efficiency is 30.7 % and the threshold wavelength of light equals 775 nm. As the losses increase, the efficiency (as well as the threshold wavelength) rapidly decreases; for a more realistic sum of losses (1 eV) the efficiency is 12.7 %. A value close to this has been obtained in Ref. [24]. It must be borne in mind that the computed value equals $K_{thr} \cdot K_{st}^{ch}$ and does not completely allow for the decrease in the photoelectrolysis efficiency due to recombination of carriers, ohmic voltage loss, and other reasons which affect the values of Y and f (see Eq. (2.12)). With the consideration of these losses the photoelectrolysis efficiency decreases down to 5 % [52]. Therefore, as follows from Fig. 31, spontaneous photoelectrolysis of water, which takes place under sunlight, is not expected to have high efficiency even when it occurs on an optimized semiconductor photoelectrode.

Thus, the first difficulty faced in realizing solar photoelectrolysis of water may be attributed to insufficient energy of the main mass of sunlight quanta for carrying out this very reaction which has a relatively high Gibbs' energy. This difficulty can be overcome by:

1. making up for the "lacking" energy of low-energy quanta; this is done by applying external voltage to the cell (the so-called photoassisted electrolysis);

---

[1] Here we recall that λ_g (nm) = 1240/E_g (eV).

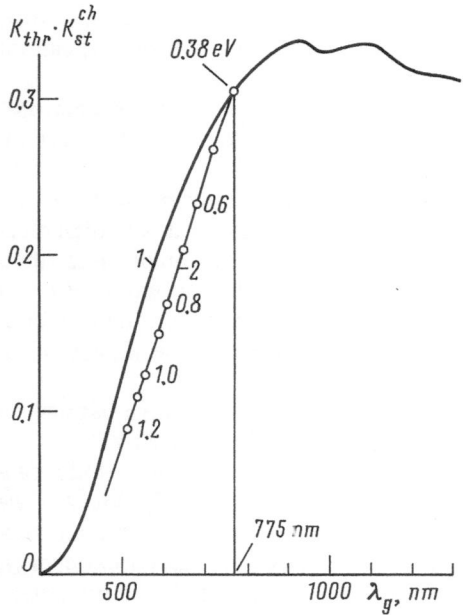

Fig. 31. Computed upper limit of $K_{thr} \cdot K_{st}^{ch}$ for photodecomposition of water [51]
1 – dependence on threshold wavelength of light at total energy losses of 0.38 eV;
2 – dependence at different values of total energy losses (shown on the curve in eV) (Reprinted by permission from Nature. Copyright © Macmillan Magazines Limited)

2. using such photoelectrochemical cells and processes in which not one but two or more light quanta are absorbed for each electron transferred in the reaction, Eq. (2.8); the energies of these quanta are summed up (the so-called two-quanta photoelectrolysis);

3. carrying out the process in two stages: first convert solar energy into electrical energy, and then (having attained the desired voltage by connecting solar cells in series) electrolyze water in a traditional electrolyzer.

Also, this difficulty can be overcome in a radical manner, i.e., by just discarding the photoelectrolysis of water and replacing it by another process with smaller Gibbs' energy (dehydrogenation of organic substances, sulfides, etc.). All these pathways will be considered in this and the following chapters.

Another difficulty entails the need of maintaining the photoelectrode Fermi level at a certain position relative to the electrochemical potential level of a corresponding reaction occurring on the electrode, or, the same, the flat band potential of the semiconductor relative to the reversible potential of the reaction (see p. 51, requirement 3). In practice, this condition is often not fulfilled. Thus, the flat band potential of one of the most-used photoanode material – titanium dioxide (rutile) – is more positive than that of the reversible hydrogen electrode. Therefore, photoelectrolysis of water on rutile does not occur spontaneously. The flat band potential could be shifted by subjecting the semiconductor to oxidation or reduction, chemisorption, etc. This method has limited posibilities, however. Because of this, the Fermi level has to be shifted by applying an external voltage, φ_{ext}, to the cell and water is but again decomposed under conditions of photoassisted electrolysis.

The energy diagram of such a cell is shown in Fig. 32. Here, n-type RuS_2 acts

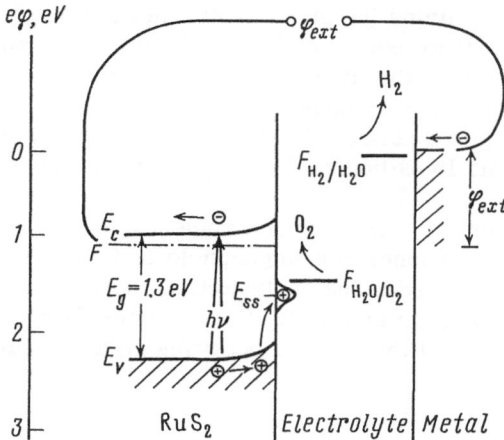

Fig. 32. Energy diagram of cell with n-type RuS_2 photoanode and metal cathode, operating under photoassisted water electrolysis conditions (reproduced from Ref. [46]) E_{ss} – surface level; in RuS_2 the valence d-band is hatched

as photoanode. The RuS_2 semiconductor is rather stable against anodic photocorrosion (photoelectrodes made of dichalcogenides of transition metals are discussed in more detail in Sect. 4.3). Its forbidden bandwidth equals about 1.3 eV; this band matches well with the solar spectrum, but is too narrow to cause decomposition of water without the application of an external voltage. Indeed, as is seen from Fig. 32, the Fermi level of RuS_2 lies much lower than F_{H_2/H_2O}. Hence, the energy of electrons in the metal electrode whose Fermi level F_{met} coincides with F_{RuS_2} when the cell is short-circuited, is inadequate to evolve hydrogen from water. The application of external voltage permits the metal-electrode Fermi level to be raised above F_{H_2/H_2O}. On illuminating the photoanode, the holes transfer to the solution (via the intermediate energy level on the RuS_2 surface shown in the figure). As a result, oxygen is evolved on the RuS_2 surface. Also electrons transfer via the external circuit to the metal cathode and then go into solution, evolving hydrogen.

The photoassisted electrolysis efficiency must be estimated by allowing for the consumption of energy from the external source of voltage. It is recommended to make use of the following formula:

$$\eta = \frac{(\Delta G/e - \varphi_{ext})\, i_{ph}}{P_1} \cdot 100\,\% \tag{3.3}$$

where P_1 is the power density of light incident on the photoelectrode. It must be stressed that φ_{ext} is the actual difference of potentials between the illuminated photoelectrode and counter-electrode when photocurrent i_{ph} flows through the cell. From Eq. (3.3) it follows that at $\varphi_{ext} > \Delta G/e$ the efficiency computed by Eq. (3.3) becomes a negative value. Physically this means that light energy is not stored during photoelectrolysis but is spent, as in photocatalysis (see Sect. 2.3), to overcome kinetic difficulties of the reaction.

Besides Eq. (3.3), other equations, for instance,

$$\eta' = \frac{\Delta G \cdot i_{ph}/e}{P_1 + \varphi_{ext} \cdot i_{ph}} \cdot 100\,\% \tag{3.4}$$

are also used in the literature for determining the photoassisted electrolysis efficiency (a summary of discussions on these equations is available in Ref. [54]). Equation (3.4) formally conforms more to the definition of efficiency given in Sect. 2.4 (cf. Eq. (2.11)), but its use is not recommended, because the efficiency computed by this equation does not give a correct idea of the amount of energy stored in the course of photoelectrolysis. In particular, the efficiency remains positive at $\varphi_{ext} > \Delta G/e$.

Besides the commonly used concept of "efficiency of a photoelectrochemical cell", sometimes one talks about the "efficiency of a single photoelectrode". The latter is conditionally determined by comparing the photoelectrode potential (φ_i) at some value of the photocurrent i_{ph} with an arbitrarily chosen "comparison potential", e.g., with reversible potential φ^0 of the reaction occurring on the photoelectrode:

$$\eta'' = \frac{|\varphi_i - \varphi^0| \, i_{ph}}{P_1} \cdot 100\,\%$$ (3.5)

or with the potential (φ_{met}) of a "good" (i.e., with low overvoltage) metal (most often, platinum) electrode in the dark when $i = i_{ph}$:

$$\eta''' = \frac{|\varphi_i - \varphi_{met}| \, i_{ph}}{P_1} \cdot 100\,\%$$ (3.6)

Equations (3.5) and (3.6) can find only restricted application as they provide an idea of the energy gain on a separately taken photoelectrode, but not in the photoelectrochemical process as a whole, because they do not take account of the losses on the counter-electrode. By way of illustration, we shall list the values of efficiency, computed in Ref. [54] with the aid of the aforementioned formulas, for water photoelectrolysis in a cell with a p-InP photocathode (the operation of this cell is considered in detail in the next section) and a platinum anode. At an external voltage of 1.25 V (i.e., at $\varphi_{ext} > \Delta G/e$) the cell efficiency determined by Eq. (3.3) is negative (−0.48 %). This is natural, because under the considered conditions the chemical energy is obtained not at the cost of the light energy but only at the expense of the external source of electricity. Equations (3.4), (3.5), and (3.6) yield large positive values: 22.7, 10.8, and 14.2 %, respectively, which hardly have any physical or practical significance.

From the aformentioned it follows that in evaluating the literature data on the photoassisted electrolysis efficiency, one must exactly specify how the efficiency was computed.

The solar energy conversion efficiency can be raised in principle by using, besides the chemical energy, the electrical energy generated by the photoelectrolysis cell. To this end, a load is brought into the external circuit. But this inevitably leads to a decrease in the photocurrent compared to the short-circuit conditions. In order to decide whether it is expedient to use the load, one must compare the total (i.e., electrical and chemical) power at the cell outlet:

$$P = P_{ch} + P_{el} = (\Delta G/e + \varphi_{ph}) \, i_{ph}$$ (3.7)

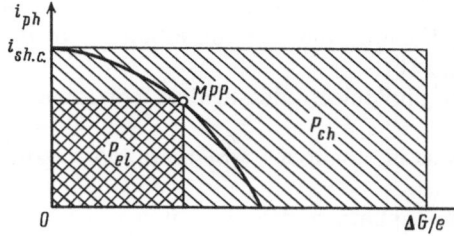

Fig. 33. Schematic for comparing the chemical and electrical power at the outlet of the photoelectrolysis cell Solid line – photocurrent-voltage curve. Shading shows the area equal to maximum chemical power; double shading – area equal to maximum electrical power (when $P_{ch}^{max} > P_{el}$)

and its maximum chemical power developed under short-circuit conditions: $P_{ch}^{max} = (\Delta G/e) i_{sh.c}$. From the physical picture of the photoelectrochemical process, considered in Sect. 3.1, it follows that the photopotential φ_{ph} determining the electrical power does not exceed the potential drop in the space charge layer, Φ_{sc}. At the same time, the maximum change in the quasi-Fermi level of minority carriers, $F - F_p$, which limits the ultimate attainable values of the Gibbs' energy of the electrochemical reaction ΔG, may be as great as the forbidden bandwidth of the semiconductor, i.e., may well exceed $e\varphi_{ph}$. Therefore, if a large energy-content product is obtained during photoelectrolysis (ΔG is large), then it is more advantageous to use the cell under short-circuit conditions. In the opposite case, the electrical energy may appreciably contribute to the cell efficiency as schematically shown in Fig. 33.

By way of example, the experimentally measured dependences of chemical and electrical components of efficiency on the cell voltage V are shown in Fig. 34 for the reaction:

$$H_2S = H_2 + S \tag{3.8}$$

(this reaction is discussed in detail in Sect. 5.4). The voltage and photocurrent (Fig. 34 a) were adjusted by suitable selection of the load resistor in the external circuit. For the reaction of Eq. (3.8), the free-energy change is not large:

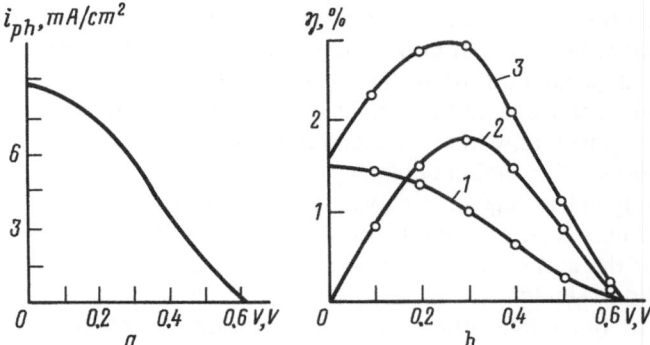

Fig. 34. Dependence of photocurrent (a) and efficiency (b) for H_2S-photodecomposition on cell voltage [55] with consideration (1) of only chemical power; (2) of only electrical power; (3) of total power

$\Delta G_{H_2S} = 0.14$ eV; therefore the use of electrical energy produced in the course of photoelectrolysis makes it possible to significantly increase the efficiency of the photoelectrochemical cell as a whole (Fig. 34b).

Attempts have been made to protect photoelectrodes against corrosion in several ways which include the use of:

1. higher oxides (TiO_2, WO_3, and others) as photoanodes, which are more stable to anodic decomposition;
2. (p-type) photocathodes instead of (n-type) photoanodes, since it is easier to select semiconductor materials more stable to cathodic decomposition than to anodic decomposition;
3. different types of protective coatings for corrosion-unstable electrodes.

These methods are discussed in detail below.

3.2 Photoassisted Electrolysis of Water

In decomposing water into hydrogen and oxygen the quanta of visible and infrared light are used to stimulate the photoelectrochemical reaction. This light contains the majority of the incident solar radiation energy, and deficient energy (up to the required amount of ca. 2 eV per quantum; see above) is supplied to the cell by making use of an additional external-source voltage. Much success in this direction has been achieved by Heller et al. [56, 57, 58]. In these works, a noticeable progress has been made by abandoning photoanodes (n-type semiconductors) on which initially main hopes were pinned, and by using photocathodes (p-type semiconductors) instead. In fact, photocathodes operate in a reducing atmosphere and therefore are much less susceptible to corrosion. That is why it has become possible to make use of semiconductor materials with optimum photoelectrical properties, such as InP ($E_g = 1.28$ eV) and Si ($E_g = 1.11$ eV), though they do not feature high corrosion resistance.

In order for the semiconductor photoelectrodes to function effectively, their surface is given a special treatment to raise its electrocatalytic activity for the hydrogen generation reaction. With this in mind, islets of platinum metals (Rh, Ru, Pt) as well as of Re are deposited on the surface, e.g., of InP. Usually, a layer of metal is electrochemically deposited on the surface of InP under illumination. After the semiconductor is etched, only traces of metal in the form of islets (clusters) of size generally of the order of several nanometer or tens of nanometer remain on the surface. Sometimes metal is deposited by vacuum deposition technique. Besides, a very thin (0.4–1 nm) layer of oxide is formed on the InP surface by mild oxidation. This layer, as the authors of Ref. [58] believe, is necessary to decrease the surface recombination velocity.

Such a treatment of the photocathode heavily increases the photocurrent (sometimes, by several orders compared to the photocathode with untreated surface) and shifts the onset potential towards more positive values. Comparing the polarization curve of the illuminated InP-photocathode with the curve drawn in

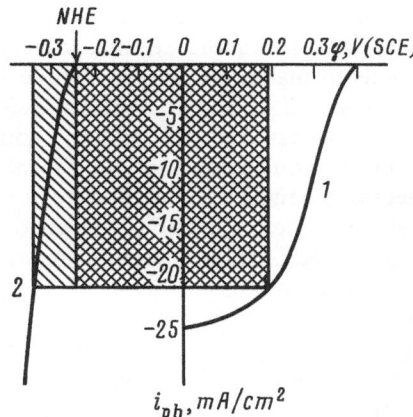

Fig. 35. Polarization curves of p-InP photocathode when illuminated (1) and of Pt cathode in the dark (2) [57]
Solution 1 M HClO$_4$, sunlight (79 mW/cm^2)
Double hatch – area proportional to photoelectrode efficiency computed by Eq. (3.5); cross hatch – by Eq. (3.6)
Potentials are given against saturated calomel electrode (Reprinted by the permission of the publisher, The Electrochemical Society, Inc.)

the dark for an electrocatalytically active metal, e.g., platinum, cathode (Fig. 35) it is not hard to see that photoevolution of hydrogen as opposed to usual electrolysis proceeds with an appreciable (up to 0.6 V) "underpotential". This is the result of utilizing light energy for electrolysis.

In the cell, chlorine or oxygen evolves at the Pt–Rh or IrO$_2$–Ir(TaO$_3$)–Ti anode made especially for the purpose, depending on the composition of the electrolyte solution (HCl, HClO$_4$, or NaOH). It should be emphasized that the reversible potential of the chlorine electrode (1.36 V) is more positive than that of the oxygen electrode (1.23 V), but the chlorine evolution overvoltage is less than that of oxygen evolution. That is why, chlorine is the main anode product in the case of hydrochloric acid solution photoelectrolysis.[2]

The issue on the efficiency of the above-desribed cells for photoassisted electrolysis demands special explanation. The authors of Refs. [56–58] have used Eqs. (3.5) and (3.6), i.e., they have computed efficiency without taking account of energy losses on the anode and, hence, total consumption of energy from the external source. The values of efficiency so obtained are listed in Table 3.1.

Table 3.1. Water photoassisted electrolysis efficiency of a cell with InP photocathode

Calculated by Eq.	Catalyst		
	Ru [56]	Rh [58]	Re [58]
(3.6)	12 %	16.2 %	14.2 %
(3.5)	8.6 %	13.3 %	11.4 %
Recalculated by Eq. (3.3)	6 %	–	–

[2] Nevertheless, the photoelectrolysis efficiency may be computed by Eq. (3.3), taking ΔG = 1.23 eV and keeping in view that the energy of obtained hydrogen can be utilized by burning it in air or reprocessing it in a hydrogen-air fuel cell, thus using not chlorine (which is spent for other purposes) but oxygen whose reserves in the atmosphere are practically inexhaustible.

From the electrode potentials given in Ref. [56] it is evident that the potential difference across the cell (i.e., the voltage φ_{ext} tapped off from the external source) equals 0.98 V when the photocurrent corresponds to maximum output chemical power. This means that about 80 % of the energy stored during photoelectrolysis ($\Delta G = 1.23$ eV) is produced at the expense of the external electric battery, and only 20 %, at the cost of solar energy. The solar energy conversion efficiency, recalculated by us by Eq. (3.3) for the InP(Ru) electrode equals 6 %, i.e., about twice less than the value given in Ref. [56]. Obviously, for other cells also the given values of efficiency are overestimated by the authors of Ref. [58] in the same proportion (compare also the values listed in Section 3.1, page 56).

The photoelectrodes were quite stable in operation: in 80 hours of continuous photoelectrolysis (in this period, 7500 C/cm² electricity passed through the photoelectrode) the electrode efficiency decreased only by 30 %. This decrease is associated, at least, partially with the electroreduction of the oxide on the InP-electrode surface (see above); it can be avoided if after every 5 minutes the cell operation is interrupted for 5–10 seconds, as under open-circuit conditions the oxide layer is spontaneously regenerated.

Silicon photocathodes with metal-catalyst clusters were tested as well [59–61]. Their efficiency was found to be somewhat less. Thus, in cells with a p-Si (Pt) photocathode, the monochromatic light (wavelength 632 nm) energy conversion efficiency was 5–6 % [62]. These electrodes are also quite stable: the photocurrent is little effected in 500 hours of continuous operation of the cell [61].

The metal clusters on other p-type semiconductor materials (GaP, GaAs, etc.) also increase the cathode photocurrent (see, for example, Ref. [60]). But the use of these semiconductors for photoassisted electrolysis, as the application of n-type photoanodes is yet less studied.[3]

Let us now see how the metal-catalysts affect the semiconductor electrode behavior. The surface of a semiconductor electrode with metal islets (Fig. 36) may be pictured as a parallel connection of junctions of two types: semiconductor/electrolyte and semiconductor/metal/electrolyte, each participating in the photoelectrochemical process. The action mechanism of the former, when a depletion layer is formed in the semiconductor, is considered in Sect. 2.1. Here, the photogenerated minority carriers (in a p-type semiconductor – electrons in the conduction band) are transferred by the electric field to the semiconductor surface, and go into solution, reducing hydrogen ions or water molecules. This process generally entails marked overvoltage. Majority carriers are transferred into the electrode bulk and then go to the cell counter-electrode via the external circuit.

The action mechanism of the latter, i.e., the semiconductor/metal/electrolyte structure, is similar. If the work function of a p-type semiconductor is more than that of a metal, then the semiconductor in contact with this metal is negatively charged. At a sufficiently large difference in the work functions, a depletion layer also appears in the semiconductor, i.e., a Schottky diode is formed. (The band bending, e Φ_{sc}, at the semiconductor/metal and semiconductor/electrolyte junctions may be different, of course.) Since the metal-layer thickness is small (e.g.,

---

[3] The operation of some oxide semiconductor photoelectrodes in photoassisted electrolysis will be considered in Sect. 3.5.2.

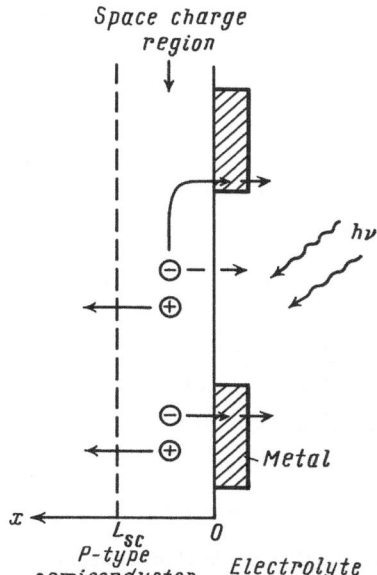

Fig. 36. Scheme of action of semiconductor photocathode with metal-catalyst islets on the surface

equals several nanometers) the layer is well transparent for light, therefore free
carriers are also photogenerated in the semiconductor under metal. These carriers
separate in the depletion layer and the electrons transfer from the semiconductor
to the metal. Further, they go from the metal to the solution. This transfer, in the
case of platinum group metals, takes place very easily, unlike the above semicon-
ductor-to-solution transfer. Thus, the semiconductor/metal/electrolyte structure
usually behaves like a solid-state (photovoltaic) cell for conversion of light energy
to electrical energy, connected in series to an ordinary electrochemical cell with
low overvoltage. If the metal layer is continuous (this case will be discussed in detail in Sect. 7.1), then the described photocurrent-generation mechanism is, of
course, the only feasible mechanism. When the layer is not continuous, i.e., the
metal is deposited as separate islets, both the considered junctions work in parallel, as mentioned earlier. Moreover, the metal islets may serve as a "drain" for
non-equilibrium electrons that appear on the portions of the semiconductor surface uncovered by the metal, as is shown in Fig. 36.

For this drain to be effective, a potential well for electrons should exist at the
InP/metal (as well as at the InP/aqueous acid solution) contact (Fig. 37 a). As
stated above, this may well be formed if the metal work function is less than that
of p-InP. However, the work functions of platinum and p-InP are practically equal
and their Fermi levels coincide (Fig. 37 b). Therefore, the p-InP/Pt contact is ohmic and Pt cannot "suck" electrons from the InP conduction band. But such a
work function for platinum (and other metals shown in Fig. 37 b) is obtained when
measured in vacuum. In an atmosphere of hydrogen (which to one or another extent
dissolves in platinum metals) the work function of platinum and, to a large degree, of rhodium decreases; it is almost equal for all three metal catalysts (Pt, Rh,

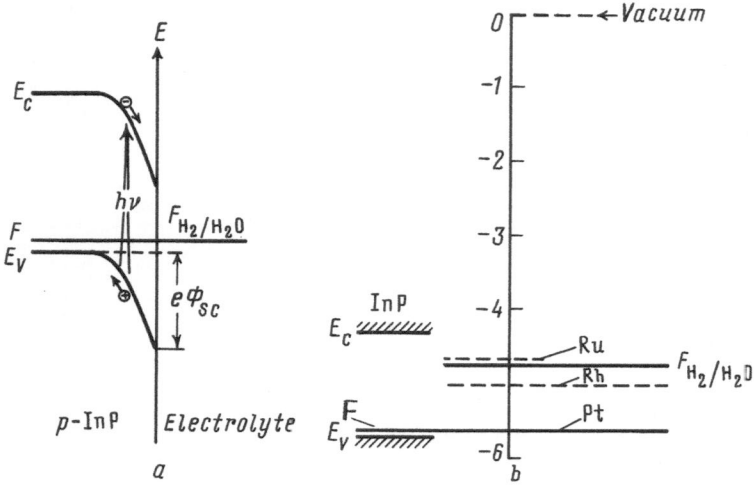

Fig. 37. To the explanation of the action of platinum metal-catalysts on p-InP photocathode and of the effect of hydrogen dissolved in them
(a) energy diagram of p-InP/aqueous acid electrolyte contact; (b) Fermi level of platinum metals in vacuum relative to InP band boundaries and to electrochemical potential F_{H_2/H_2O} in aqueous solution [58] (Reprinted with permission from Journal of American Chemical Society. Copyright (1982) American Chemical Society)

and Ru). Moreover, the Fermi level for these metals when saturated with hydrogen is very close to the electrochemical potential level of the hydrogen electrode, F_{H_2/H_2O}, in aqueous solution [63, 64]. This means that, first, the Schottky barrier height at the InP/hydrogen-saturated metal junction depends not so much on the properties of a particular metal rather than the properties, as suggested by the authors of Ref. [63], of "metallic hydrogen", and, second, this height is rather large – it is almost the same as at the p-InP/aqueous acid solution contact (because an equilibrium hydrogen electrode potential is established on the InP electrode due to the adsorption of hydrogen on the electrode surface).

This is what happens in a photoelectrochemical cell; thanks to the sufficient depth of the potential well for electrons, the platinum metal islets readily enhance the hydrogen-evolution process. As the desired band bending does not exist initially but appears during photoelectrolysis when hydrogen evolves on the electrode and penetrates deep into the metal, optimal characteristics of the photocathode are obtained not immediately but after a certain induction period (as this often happens in autocatalytic processes). On bringing the electrode into an oxygen atmosphere it loses its activity which is restored as soon as the metal again gets saturated with the released hydrogen.

For the electrons photogenerated on the free portions of the semiconductor, the proportion of their transfer into solution directly or via metal islets depends on the distance between the islets, diffusion length of electrons and, of course, on the ratio of rate constants of hydrogen ions (or water molecules) reduction reactions on the semiconductor and on the metal.

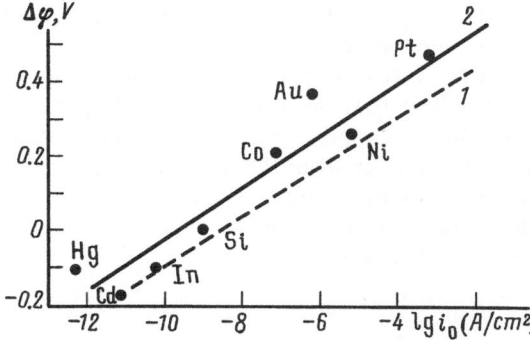

Fig. 38. Dependence of potential shift of p-type Si photocathode under illumination (1) and of n-type Si cathode in the dark (2) (at a current density of 5 mA/cm^2), owing to deposition of microislets of metals, on the logarithm of exchange current of "hydrogen" reaction on electrodes made of the same metals [59]

Sometimes, the metal (in case of discontinuous layer) directly acts as catalyst for the hydrogen evolution reaction. This becomes clearly evident on comparing the hydrogen evolution rate on silicon electrodes with islets of metals and on these very metals taken as cathodes. In fact, as shown in Fig. 38, the acceleration of hydrogen photoevolution from aqueous solution on p-type silicon electrodes, as well as of dark hydrogen evolution on n-type silicon with metal islets (estimated as "positive" potential shift $\Delta\varphi$ of the Si electrode overlaid with a metal-catalyst relative to the electrode without a catalyst) is a linear function of the logarithm of the exchange current for the hydrogen electrode reaction on these metals. At the same time, no direct relationship between the catalytic action of metals and their work function was noticed. Thus, the use of metals having large work function does not decrease the Si electrode photopotential, though the Schottky barrier height at the silicon/metal junction will decrease in this case, which would impair the photoelectric properties of the junction. Consequently, the function of metal microdeposits is mainly to speed up the evolution of hydrogen [59–61]; photogeneration and separation of charges take place mostly on the electrode surface not covered with the metal. (How large the contribution will be of photogeneration of carriers strictly under metal spots to the total photocurrent depends, in the general case, on the relative fraction of the surface covered with metal, the transparence of the metal, etc.)

Note that if the metal-particle size is very small (tenth fractions of a nanometer) then the appearance of quantum-size effects might be expected in their electrochemical behavior [65].

Along with the metal-catalysts deposited on the semiconductor surface, homogeneous catalysts (mediators) may also be used to accelerate hydrogen evolution from aqueous solutions. Homogeneous catalysts are redox systems like Cr(II)/Cr(III), V(II)/V(III), and some heteropolyacids, for example, $H_8(SiMo_{12}O_{42})$ [66]. The idea of using such catalysts is that, though the equilibrium potential of these systems is slightly more negative than the reversible hydrogen electrode potential, the overvoltage of their reduction at the semiconductor photocathode may be much less than the hydrogen evolution overvoltage. The reduced form of a mediator obtained in the photoelectrochemical reaction is capable of liberating hydrogen from water in the course of the chemical reaction occurring in the solution bulk. In this reaction, the oxidized form is regenerated which again enters into the reduction reaction at the photocathode, and so on.

Thus, the cells with photocathodes are very efficient under conditions of photoassisted electrolysis of water. Nevertheless, they are yet not ready for practical use. The main difficulties encountered in their application are: insufficient stability and very large consumption of electric energy from the auxiliary source. This compels, along with further development of the photoassisted electrolysis technique, to find alternative routes for photodecomposition of water, such as two-quantum photoelectrolysis and two-stage processes.

3.3 Two-Quantum Photoelectrolysis of Water

Photoelectrochemical cells for two-quantum photoelectrolysis bear two photosensitive interfaces which can be disposed as follows:
a) two photoelectrodes – anode and cathode – in the same cell;
b) two photoelectrodes in two separate compartments of the cell, connected to each other through a chemical substance – a charge carrier;
c) photoelectrode with inner p-n-junction (or isotypic heterojunction) is disposed in such a manner that the photovoltage developed across this junction upon illuminating the electrode will accelerate the photoelectrochemical reaction occurring at the photoelectrode/electrolyte interface.

In all these cases the electrochemical potentials generated at both photosensitive interfaces in the cell are summed up. In cases (a) and (c) the electrochemical potentials of electrons in the semiconductor(s) add up; in case (b) the chemical potentials of two redox systems in the solutions, present in the cell compartments, are added together.

3.3.1 Cells with Two Photoelectrodes

Such a cell (Fig. 39) has a n-type semiconductor photoanode and a p-type semiconductor photocathode. The areas of these photoelectrodes are so taken that upon illumination their photocurrents (depending on the light absorption coefficient, quantum yield, and recombination velocity) are equal in absolute value.

As is seen from Fig. 39, it is essential that the conduction band edge of the p-type semiconductor is above the electrochemical potential level of the water reduction reaction, and the valence band edge of the n-type semiconductor below the electrochemical potential level of the water oxidation reaction. These differences must be optimized in magnitude so as to avoid unnecessary losses. All this imposes very exacting requirements on the flat band potential and the forbidden bandwidth of both semiconductor materials; in practice, it is not easy to meet these requirements.

The theoretical (limiting) efficiency of a cell with two photoelectrodes was calculated [52] by making the following (very optimistic) assumptions: "cathode over-

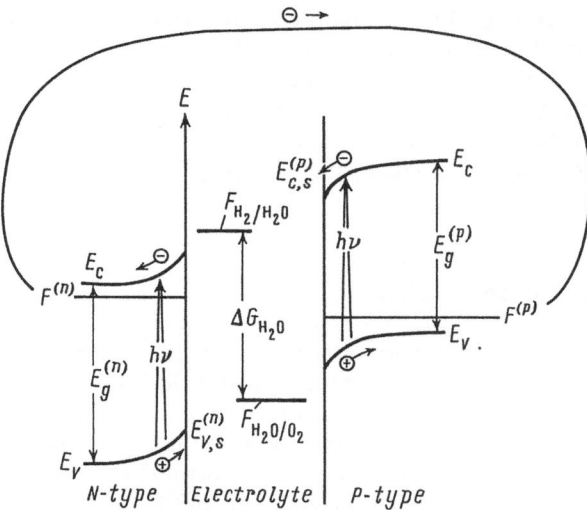

Fig. 39. Energy diagram of cell with two photoelectrodes

voltage" $(E_{C,s}^{(p)} - F_{H_2/H_2O})/e = 0.05$ V; "anode overvoltage" $(F_{H_2O/O_2} - E_{V,s}^{(n)})/e = 0.4$ V. Other losses that limit the values of Y and f were allowed for in the following manner: photocurrent-voltage curves of a photoelectrochemical cell were taken as the analogous curves of the best solid-state solar cells (made of monocrystalline silicon). The computed dependence of the efficiency on the forbidden bandwidth (which was taken the same for both semiconductors) is shown in Fig. 40. Maximum water photoelectrolysis efficiency (9.8 %) is attained at $E_g = 1.75$ eV. (Note that the semiconductor electrode with a forbidden band of this width is alone unable to cause spontaneous photodecomposition of water.)

Is it possible to make such a photoelectrochemical cell by allowing for numerous and very exacting requirements placed on its electrodes? For a long time all attempts of making such a cell were unsuccessful. Even though the cells with, for example, a n-type TiO_2 or $SrTiO_3$ photoanode with a p-type GaP or CdTe photocathode can cause spontaneous photodecomposition of water, they do not function very effectively (see Sect. 9.3.3 in Ref. [1]). Thus, the cell with ceramic photoelectrodes made of Fe_2O_3 (doping with 2 % silicon enables n-type material to be ob-

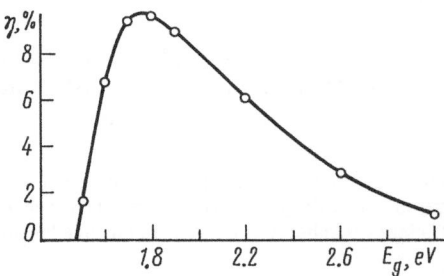

Fig. 40. Computed dependence of water photoelectrolysis efficiency in cell with two photoelectrodes on forbidden bandwidth [52] (Reprinted by the permission of the publisher, The Electrochemical Society, Inc.)

tained for the photoanode, and doping with 5 % Mg, p-type material for the photo-cathode) features good stability of photocurrent. However, the water photoelectro-lysis efficiency does not exceed 0.1 % [67], i.e., is much less than in better cells with one photoelectrode.[4] But a similar cell (with a n-type $MoSe_2$ or WSe_2 photo-anode and a p-type InP photocathode) operates more effectively (see Sect. 4.1) for photoelectrolysis of substances other than water, for example, hydrobromic and hydroiodic acids.

3.3.2 Two-Chamber Cells for Photoelectrolysis

This method is based on splitting the complete water decomposition process into two stages, each occurring in a separate photoelectrochemical cell. Both cells are combined together via some redox system whose oxidized and reduced compo-nents (Ox and Red) act as charge carriers; these components are not consumed in the course of the net process, they only ensure "series connection" of chemical potentials developed in both cells (Sect. 9.3.3 in Ref. [1]; [69]). Such a unit (Fig. 41) actually simulates the combination of two photosystems in the well-known Z-shaped reaction scheme of photosynthesis in green plants.

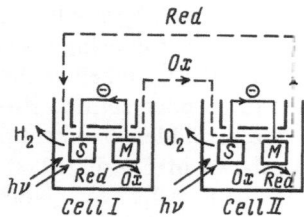

Fig. 41. Schematic of a two-chamber cell for two-quantum water photodecomposition
S - semiconductor photoelectrode; M - metal electrode; diaphragms that separate the electrodes are not shown

The problem resides in selecting a charge carrier system with a reversible po-tential such that splitting of water into hydrogen and oxygen is effected in two subsequent steps. For either step the Gibbs' energy change is less than ΔG_{H_2O} = 1.23 eV. In cell I, water gets reduced to H_2 on the photocathode and, on the me-tal anode, the reduced component, Red, gets oxidized to Ox. Further, Ox goes to cell II where water oxidizes to O_2 on the photoanode and Ox reduces to Red on the metal cathode; the reduced form returns to cell I. Thus, at the cell outlet the water decomposition products are obtained as before, but this process now pro-ceeds with the absorption of two light quanta per electron transferred in the reac-tion, Eq. (2.8).

No suitable charge carrier has yet been found for the water-splitting reaction. The following systems are proposed for CO_2 photoreduction: I^-/I_2, S^{2-}/S_2^{2-}, and methylviologen [70].

---

[4] Very recently the authors of Ref. [68] have reported about a cell with p-type InP and n-type GaAs photoelectrodes (the surface of GaAs is covered with thin protective-catalytic layers of Pt and MnO_2) in which the water photoelectrolysis efficiency reaches 8 %. But this report needs con-firmation.

In a two-chamber cell the limiting (theoretical) photoelectrolysis efficiency should obviously be the same as in the above-considered cell with two photoelectrodes. On the whole, this pathway of photoelectrolysis is yet less studied.

3.3.3 p-n-Junction and Heterojunction Photoelectrodes

If a semiconductor photoelectrode has deep-seated interphase boundaries featuring photosensitivity and all of them, along with the electrode/electrolyte interface, are illuminated simultaneously, then the photopotentials and/or photocurrents appearing on them are summed up, encouraging more intense occurrence of the photoelectrochemical reaction on the electrode surface. Thus, this reaction becomes essentially a two- (or multi-)quantum reaction.

A necessary condition for such a complex electrode to function is that the photoelectrically active light must penetrate into the electrode interior. This will happen if the outer layer is sufficiently transparent, e.g., if it is made of a semiconductor material with a wider bandgap than the inner region. Then the relatively long-wave light passes through the outer layer practically with no attenuation. On reaching the narrow-bandgap semiconductor, the light is absorbed by it, generating non-equilibrium carriers. Short-wave light is effectively absorbed by the outer layer and the non-equilibrium carriers formed therein directly participate in the photoelectrochemical reaction on the electrode surface. Thus, double- (or multi-) layer semiconductor structures made of different materials (generally known as heterojunctions) enable the entire spectrum of sunlight to be effectively used. Solid-state photovoltaic cells functioning according to this principle are called cascade cells (see, for example, Ref. [71]).

The simplest and one of the initially proposed photoelectrodes of this type is a usual n-p^+-junction silicon solar cell on the outer (p^+) face of which a polycrystalline TiO_2 layer of thickness of about 500 nm is deposited (Fig. 42 a). Under the action of ultraviolet light, holes are generated in TiO_2 ($E_g = 3$ eV) which oxidize water to oxygen. The infrared light, without being absorbed in TiO_2, reaches silicon ($E_g = 1.11$ eV) and the photovoltage developed at the n-p^+-junction shifts the anode polarization curve of TiO_2 (relative to the curve for a TiO_2 electrode on the silicon substrate, but without a p-n-junction) towards more negative potentials, as is shown in Fig. 42 b. Upon illuminating the cell with a described photoanode and a platinum cathode, water spontaneously decomposes into hydrogen and oxygen. Here, the silicon p-n-junction acts as an external voltage source relative to the ti-

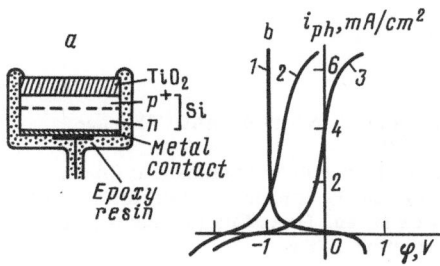

Fig. 42. (a) Schematic of n-p^+ Si/TiO_2 electrode; (b) current-voltage curves in 0.1 M NaOH
1 – cathodic curve on Pt; 2 – anodic curve on n-p^+-Si/TiO_2; 3 – anodic curve on n-Si/TiO_2 [72]

tanium oxide photoanode which, therefore, functions as if under photoassisted electrolysis conditions. It must be borne in mind that the flat band potential of TiO_2 (rutile) is more positive than the reversible hydrogen potential; therefore, spontaneous photodecomposition of water in an ordinary cell with a rutile anode is not possible (for details, see Sect. 3.5.2).

The considered type of photoelectrode is a combination of a solid-state photovoltaic cell for conversion of light energy to electrical energy and a photoelectrochemical cell for conversion of light energy to chemical energy, placed in the circuit in series. Such photoelectrodes are sometimes called "tandem" electrodes. The photovoltaic part of the tandem can be a:

1. p-n-junction in a constant-composition semiconductor (the so-called homojunction, see above);
2. junction between two different semiconductors (the so-called heterojunction), which in its turn may be
    a) isotypic (both semiconductors are of the same conductivity type) or
    b) anisotypic (semiconductors with different types of conductivity).

An example of an isotypic heterojunction is the n-type GaAs/n-type CdS photoanode [73] (which was indeed used for photoelectrolysis not of water but of sulfides). Its energy diagram is shown in Fig. 43 a. As follows from the mutual disposition of valence band boundaries in GaAs and CdS, a potential barrier for holes exists at the interface between the semiconductors. Therefore, the holes photogenerated in the GaAs valence band cannot pass to the CdS valence band and reach the electrolyte; they recombine with the conduction band photoelectrons generated in the CdS and thereby carry the photocurrent through the heterojunction. The GaAs photoelectrons transfer across the ohmic (metal) contact and then go via the external circuit to the cell counter-electrode. Only the holes generated in CdS directly participate in the photoelectrochemical reaction on the electrode/electrolyte interface. The photovoltage developed across the heterojunction causes

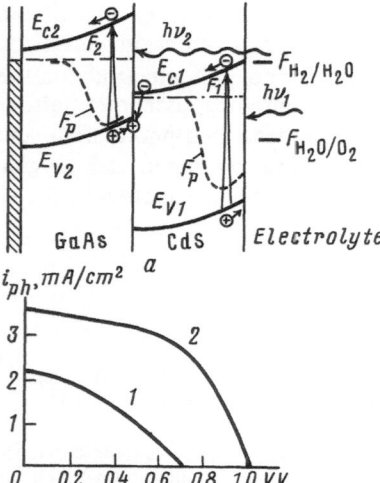

Fig. 43. Photoanode with n-GaAs/CdS heterojunction [73]
(a) Energy diagram upon illumination; (b) current-voltage curves of photoelectrochemical cell with a CdS (1) and GaAs/CdS (2) photoanode in sulfide solution. $v_1 > v_2$

the photocurrent and photovoltage of the electrode to increase compared to an ordinary CdS electrode (Fig. 43 b).

The situation discussed for the GaAs/CdS junction is typical for the majority of photoelectrodes with heterojunctions. Because the inner layer is illuminated through the outer, these layers should have markedly different forbidden bandwidths. (The wide-band material is disposed on the exterior, contacting with the electrolyte.) Hence, the heterojunction itself necessarily serves as a potential barrier for the movement of free carriers from the photoelectrode interior to its surface (we shall return to this situation in Sect. 6.2.1); therefore, in the photoelectrode with a heterojunction these are the photovoltages but not the photocurrents of individual interfaces that are summed up.

Compared to the cell with two photoelectrodes (Sect. 3.3.1) and to the two-chamber cell (Sect. 3.3.2) in which both photoelectrodes are illuminated independently, here a gain in the photoelectrolysis efficiency must be obtained owing to better utilization of sunlight. The limiting theoretical efficiency of cells with a tandem photoelectrode, computed in Ref. [52] by the method described on page 65, is represented in Fig. 44 as a function of the forbidden bandwidth of both semiconductors. It reaches 16.7 % (compared to 9.8 % for cells with two photoelectrodes) for the forbidden bandwidth of the inner ("photovoltaic"), 1.4 eV, and outer ("photoelectrode") material, 1.95 eV. The semiconductors GaAs and GaP, for example, could roughly form such a pair.

Various kinds of heterostructure photoelectrodes are described in the literature. These heterostructures are mainly isotypic: n-Si/Fe$_2$O$_3$ [74] with an experimentally obtained water photoelectrolysis efficiency of up to 1.6 %; n-Si/TiO$_2$, n-Si/Bi$_2$O$_3$, WO$_3$/TiO$_2$ [75], and anisotypic: p-Si/TiO$_2$ [26], n$^+$-p-Si/NiO(OH) with a very high photocurrent [76], amorphous Si/p-Si [77], and others. In the majority of cases their manufacturing technique was not optimized and therefore their photoelectrochemical characteristics are far from those predicted theoretically. The main difficulties are associated with the need of choosing the photosensitivity of different parts of the electrode precisely so that their photocurrents (determined by the amount of light that reaches each part, absorption coefficient, height of potential barriers, resistance, etc.) are the same, because when connected in series the "weakest" portion (i.e., with least photocurrent) limits the efficiency of the sys-

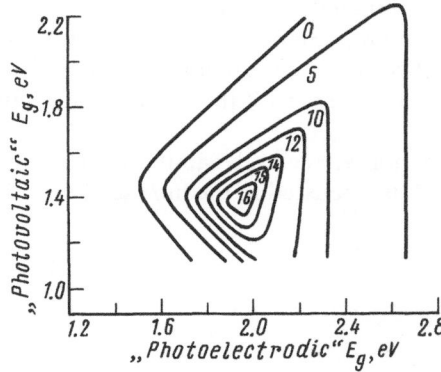

Fig. 44. Water photoelectrolysis efficiency in cell with tandem photoelectrode versus forbidden bandwidth of semiconductors constituting the electrode [52]
Limiting efficiencies (in %) are shown on curves (Reprinted by the permission of the publisher, The Electrochemical Society, Inc.)

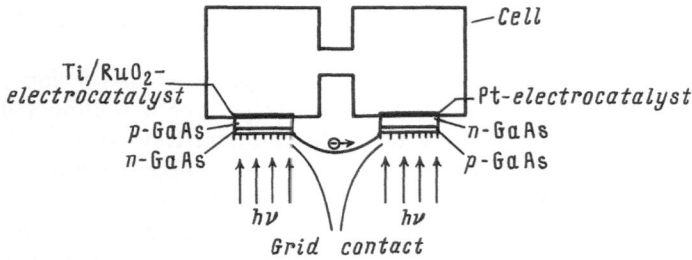

Fig. 45. Cell for photoelectrolysis of water with two p-n-junction photoelectrodes [78] in 5 *M* H₂SO₄ solution

tem as a whole. It is difficult also to avoid significant recombination of carriers on the defects which usually appear due to a mismatch of the crystal lattice of two different semiconductors at a heterojunction.

In evaluating the prospects of "tandem" electrodes one must take into account that manufacturing of such electrodes calls for a comparatively sophisticated technology (for deposition of multi-layer structures and for making p-n-junctions). At first glance this deprives such electrodes of the main advantage of photoelectrochemical cells over solid-state photovoltaic cells: they are simple to manufacture and, hence, cheap. Nevertheless, thanks to more complete utilization of sunlight and to the possibility of using non-traditional semiconductor materials as electrodes, this variant of photoelectrolysis may be regarded as a prospective variant.

In closing, we shall consider a cell for photoelectrolysis of water [78] which has been made by integrating the principles described in this and earlier sections. It has two GaAs photoelectrodes each having a p-n-junction (Fig. 45). These electrodes are glued to the orifices in the walls of the cell filled with 5 *M* H₂SO₄ solution. The electrode materials proper: Pt (for the evolution of hydrogen) and Ti/RuO₂ (for the evolution of oxygen) are coated directly on GaAs. Under illumination, photoelectrolysis of water takes place and the efficiency equals 7.8 %; here, no external voltage is required. In another variant of this cell [79] GaAs is replaced by amorphous silicon (three thin-film p-i-n-photocells connected in series whose total thickness, as low as 700 nm, enables them to be illuminated all the way through). Because of low efficiency of the amorphous silicon cell, the water photoelectrolysis efficiency is also small (2.8 %). But it must be borne in mind that the cost of amorphous silicon solar cells tends to decrease rapidly, which makes such cells economically attractive even if their efficiency is relatively small.

The cells described at the end of this section are actually similar to those for the two-stage photoelectrolysis of water which is discussed in the next section.

3.4 Two-Stage Conversion of Solar Energy: Generation of Electrical Energy Followed by Electrolysis of Water

In a traditional cell for photoelectrolysis of water one generally encounters a difficult problem of combining good photoelectrical and electrocatalytic characteristics in a semiconductor material. Upon replacing a single photoelectrolysis cell with a solid-state solar cell and an electrolyzer, the production of electrical power and electrochemical splitting of water, combined in a cell for photoelectrolysis, take place in two special-purpose facilities. This makes it possible to attain the desired voltage for splitting water by connecting several solar cells in series, and allows much room for manipulation in choosing the photoelectrical and electrocatalytic characteristics of the material used.

Thus, the following are the advantages of the two-stage hydrogen production method:
1. the conversion efficiency can be increased by adopting optimal materials for the photoelectrical and electrochemical parts of the plant;
2. corrosion peculiar to cells with semiconductor electrodes can be eliminated.

The disadvantages of this method are:
1. some (though small) increase in losses upon transmitting current over the wires connecting the two parts of the plant;
2. the impossibility of using "hot" carriers for conducting the photoelectrochemical reaction (see Sect. 2.1). (Here, it must be mentioned that no one has so far succeeded in using "hot" carriers even in photoelectrolysis cells.)

The proposed method may not be regarded as an alternative to the purely photoelectrochemical solar energy conversion method. On the contrary, by allowing for a much higher degree of development of the technological aspects of both solid-state solar cells and electrolyzers, compared to photoelectrochemical cells, one may think that in the near future this two-stage energy-conversion method, despite the high cost of solar cells, may find certain recognition while the comparatively more distant future is for the photoelectrochemical solar energy conversion method.

3.4.1 "Solar Cell + Electrolyzer" Plant

Two-stage production of hydrogen at the expense of solar energy essentially involves the combination of a solar cell and an electrolyzer. In the below considered plants, commercial monocrystalline silicon solar cells as well as commercial electrolyzers are used. The latter are, as a rule, employed for splitting water (though, of course, other substances can also be electrolyzed). The solar cell may be connected to the electrolyzer directly or via a matching device (transducer, see below). The complete plant (Fig. 46) should comprise, of course, devices for storage of obtained hydrogen and its later use, i.e., for generating electrical energy (fuel cell) and/or heat (hydrogen burning torch) from hydrogen. Here, we shall restrict ourselves to a discussion of the interaction of a solar cell with an electrolyzer.

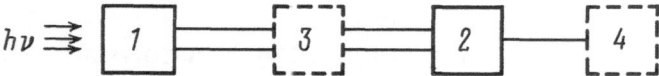

Fig. 46. Schematic of plant for two-stage conversion of solar energy
1 – solar cell; 2 – electrolyzer; 3 – matching device; 4 – devices for storage and utilization of hydrogen
Units not discussed in detail are shown by dashed lines

The efficiency of a plant is quantitatively characterized by the so-called "sun-to-hydrogen" efficiency:

$$\eta_{S\text{-}H} = \eta_s \cdot \eta_e \tag{3.9}$$

where η_s is the efficiency of the solar cell and η_e is the "efficiency of the electrolyzer". The latter takes account of energy losses both in the electrolysis process proper and due to poor matching of these two parts of the plant.

Because η_s is in any case preset by the chosen solar cell, the task of making an efficient plant amounts to increasing η_e in every possible way. To this end, besides using a high quality electrolyzer, the characteristics of the electrolyzer and solar cell should be well matched.

The operation point of the system as a whole conforms to the intersection of current-voltage characteristics of the solar cell and the electrolyzer (Fig. 47), i.e., to the conditions[5]:

$$I = I_s = I_e \quad \text{and} \quad V = V_s = V_e \tag{3.10}$$

where I is the current; V is the voltage (the subscripts s and e refer respectively to the solar cell and electrolyzer).

The solar cell operates most effectively at the point of maximum power (Fig. 47; cf. Fig. 28). Therefore, the current-voltage performance characteristic of the electrolyzer should exactly pass through this point. This aim is easily attained because the solar cell and the electrolyzer usually have a modular structure, i.e., consist of a number of identical parts (modules). By connecting these modules in series circuits over N_s and N_e of the modules and combining these circuits in parallel, the current-voltage curves can be arbitrarily changed within the given dimensions (areas) of both components of the plant, maintaining the power constant. This matching scheme is shown in Fig. 47a. Dividing the total area S_s of the solar cell into 2, 4, and more parts, we get, instead of curve 1, curves 2, 3, etc., and their points of maximum power lie on the hyperbola $P_{max} = (I\,V)_{max} = \text{const}$. The same can be done with the current-voltage characteristic of the electrolyzer (this is not shown in Fig. 47). Adjustment of N_s and N_e is continued until the obtained curves intersect at the solar cell maximum power point.

Even after choosing the values of N_s and N_e (as shown in Fig. 47a) which are optimum for a certain preset illumination intensity, the optimization problem

---

[5] Here, for simplicity, the ohmic voltage drop in the connecting wires has not been allowed for.

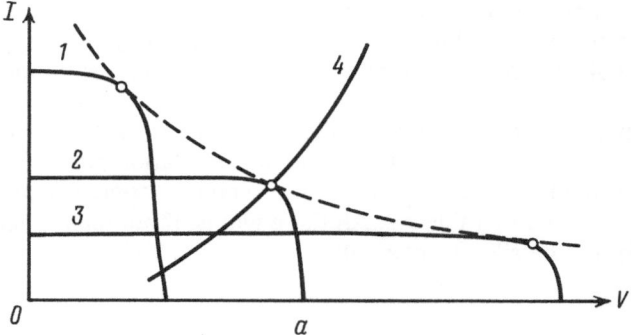

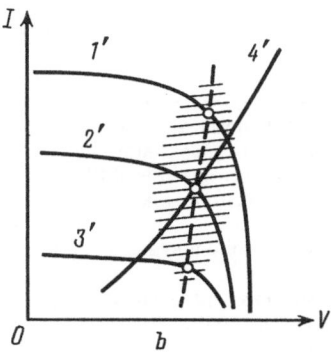

Fig. 47. Matching of current-voltage characteristics of solar cell and electrolyzer
(a) at constant light power density: 1, 2, 3 - characteristics of solar cell at $N_s = 1, 2$, and 4 (at
$S_s = $ const); 4 - characteristic of electrolyzer ($N_e = 1$). o - maximum power point (MPP);
dashed line shows the hyperbola of maximum cell power at the given light power density; (b)
at varying light power density: 1′, 2′, 3′ - characteristics of solar cell; dashed line - geometric
place of MPP; hatched region - variations of MPP in the most probable limits of variation of
light power density and temperature; 4′ - characteristic of electrolyzer

cannot be considered to be conclusively solved. In fact, the illumination intensity
varies depending on the daytime, season (these are periodic variations) and cloudiness (random variations). The geometric place of maximum-power point on the
curves corresponding to different illumination intensities is usually an "almost
vertical" line in Fig. 47b. Account should also be taken of the temperature effect
on the solar cell characteristic, because temperature varies throughout the day and
depending on the season. This effect would have been expressed by an "almost
horizontal" line (not shown in Fig. 47b). As a result, the values of voltage V_{max} and
current I_{max} conforming to the cell maximum power, vary within a certain region
(hatched in Fig. 47b). For every illumination intensity and temperature there
exists a pair of optimal values of N_s and N_e. And if these values are rigidly speci-
fied, then a disbalance (since the operation point of the plant does not coincide
with the maximum-power point of the solar cell) appears from time to time. The
disbalance somehow decreases the energy conversion efficiency.

For the experimental plants described below in this section, N_s and N_e have been taken empirically. In the next section we shall present a specially developed method for optimizing the plant design (i.e., for the theoretical calculation of optimal values of N_s and N_e).

In the past decade many "solar cell + electrolyzer" plants of various sizes ranging from laboratory scale (capacity several watts) to demonstration plants (capacity several tens of kilowatts) have been developed in several countries. The parameters of some of them are listed in Table 3.2. All these plants, if not mentioned otherwise, utilize monocrystalline silicon solar cells.

Table 3.2. Characteristics of "solar cell + electrolyzer" plants

Year of manufacture	Output P_{max}, kW	Efficiency, %			Remarks	Ref.
		η_s	η_e	η_{S-H} [a]		
1982	1	14	50	7	Use is made of a heliostat to track the Sun and of a solar concentrator (Fresnel lens). Electrolyzer is of filter-press type. Electronic matching device	[80]
1982	0.004 0.1 [b]	6–7	70	4–5	Electrolyzer with a solid polymer electrolyte	[81]
1982	8	–	–	–	The plant design envisages storage of hydrogen. Output 7000 kWh/year. Connection – direct or via a matching device on operational amplifiers	[82]
1985	0.006	9	80	7	Electrolyzer with a solid polymer electrolyte	[83]
1986	0.1	–	–	–	Filter-press type electrolyzer; electrical battery for smoothing out current fluctuations in the electrolyzer, occuring due to variation in the illumination intensity	[84]
1986	1	–	–	–	Diaphragm-type electrolyzer. Storage of hydrogen	[85]
1986	1	6.1	65	3.7–4	Filter-press type electrolyzer	[86]
1986	10	–	–	–	Mobile plant installed on the trailer chassis. Collapsible solar array. Electrolyzer with a solid polymer electrolyte. There are water storage tanks and a water purification (deionization) unit	[87]
Designed in 1986	100; 2; 10	–	–	–	The plants designed under the HYSOLAR program for operation in Riyadh and Jidda (Saudi Arabia) and Stuttgart (Federal Republic of Germany). Three different types of matching devices are used	[88]

[a] Referred to heat content of hydrogen $\Delta H = 285$ kJ/mol.
[b] The solar cell is made of polycrystalline silicon.

Here, two types of electrolyzers have been used: alkaline (usually of filter-press type) and more sophisticated ones with a solid-polymer electrolyte. The latter can operate at a relatively low voltage (1.55-1.60 V) and generally call for heating. (They can be warmed up, say, by the heat dissipated in the solar cell.)

By way of illustration, we shall describe the plant discussed in Ref. [85] in somewhat greater detail. The solar cell is assembled from 120 modules each having a peak power of 8.2 W; these are combined in two groups of 60 serially connected modules. At an incident light power density of 700 W/m^2, the cell yields a current of 20 A at 27 V. This cell is connected directly to the electrolyzer consisting of two modules each having 8 serially connected elements. The alkaline electrolyzer contains an asbestos diaphragm and metalloceramic electrodes. Its operating voltage is 1.75-1.8 V, current density 240 mA/cm^2, current efficiency 85 % (the obtained hydrogen is 99.8 % pure). The capacity of a metal-hydride-type hydrogen storage unit is 58 kWh; its output (during prolonged usuage) is 4.5 kW.

The designed plant [82] incorporates a complete cycle of conversion of solar energy, first, into electrical energy and then into chemical energy of hydrogen which, after storage, is again converted into electrical energy using a fuel cell. The estimation of total losses in all these stages reveals that at $\eta_s = 15\%$ the output electrical energy is obtained with an efficiency of 4.5 %.

We shall give some attention to a plant [84] which, along with a solar cell and an electrolyzer, contains also an auxiliary electrical battery that gets charged when the light intensity is maximum (i.e., in midday) by drawing current from the electrolyzer. At periods of low intensity (e.g., in the morning and evening) the battery, on the contrary, gets discharged on the electrolyzer. Thanks to this, the hydrogen evolution rate is less prone to fluctuations in light intensity. Generally speaking, batteries can be used also for storing solar energy over a longer period of time. It is not unlikely that in some cases this method of solar energy storage will be preferred to storage of energy in the form of hydrogen.

The studies made in different countries on plants of different design and output enable us to draw the following conclusions:

1. The "solar cell + electrolyzer" plants readily lend themselves to mathematical simulation. The methodology for such a simulation has been worked out [89-91]. Comparison of computed characteristics of a plant as a whole, such as efficiency η_{s-H}, current and integral output, with the experimentally measured values displaces high precision of calculations (Fig. 48). At the same time, it has been shown that the integral production of hydrogen (say, for one year) is little effected by short-time fluctuations in light intensity.

2. Usually the losses by the above-mentioned mismatch of solar cell and electrolyzer, associated with light intensity fluctuations, are small: 6 to 9 % of the transformed power [87] (cf. also Fig. 48, curve 2). This casts doubt on the expediency of using special electronic matching devices (transducers) because the power losses in these devices are of the same order of magnitude.

3. The economic calculations made in Refs. [80, 82, 86, 90, 92] show that the cost of hydrogen obtained at the expense of solar energy is still very high. Though the calculations are based on several distinct (in different countries) principles, nevertheless, they have yielded relatively well-coinciding results: 0.5-1.4 US dollars for 1 kWh of energy contained in hydrogen (referred to heat

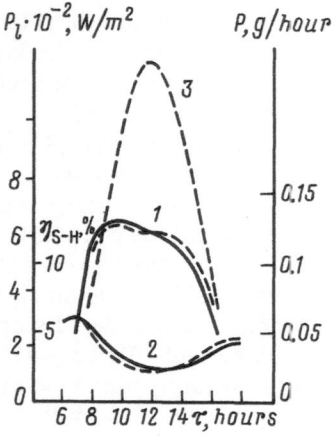

Fig. 48. Comparison of computed (solid lines) and measured (dashed lines) characteristics of "solar cell + electrolyzer" plant throughout the day [89]
1 - hydrogen production rate (P); 2 - sun-to-hydrogen efficiency (η_{S-H}); 3 - incident light power density (P_l) curve as a function of daytime

content). This is about 1–1.5 orders of magnitude more than the cost of hydrogen obtained from traditional sources. But there are sound hopes for further decrease in the cost of solar cells, therefore the just now mentioned cost of hydrogen does not seem to be catastrophically high. Even at this price the cost of autonomous plants in remote areas where there are no other sources of energy, may become expedient within the next few years.

From these calculations it also follows that the value of the electrolyzer comes to 15–25 % of the total cost of the plant, which is small compared to the cost of the solar cells. Therefore, it is beneficial to increase the relative dimensions of the electrolyzer so that it will operate at a smaller current density. This is how the losses due to overvoltage in the electrolyzer can be decreased and η_e increased.

3.4.2 Optimization of "Solar Cell + Electrolyzer" Plants [91]

As the technology both of solar cells and electrolyzers taken separately is sufficiently well developed, the task of their matching with each other (i.e., the optimization of the plant as a whole) is placed in the forefront. Depending on the requirements, this problem can be solved for different objective functions (output, efficiency, cost of unit product, etc.). The proper choice of the objective function plays the most important role in compiling the optimization recommendations. In the below-described optimization method, the optimization conditions for a plant as a whole conforms to the operation of the solar cell under maximum power conditions.

Basic relationships. The current-voltage characteristic of a solar cell having total working area S_s at a certain irradiation intensity J_0 can be expressed as:

$$i_s = S_s f_0 (J_0, v) \tag{3.11}$$

where i is the current; v is the voltage; $f_0(J_0, v)$ is the specific (i.e., per cm² of the cell area) characteristic at the irradiation intensity J_0. The current-voltage characteristic of an electrolyzer with total electrode area S_e can be expressed as:

$$i_e = S_e g_0(v) \tag{3.12}$$

where $g_0(v)$ is the specific (i.e., per cm^2 of the electrode area) characteristic of the electrolyzer.

Dividing the total area of the solar cell into N_s similar series-connected unit solar cells (modules) gives:

$$I_s = \frac{S_s}{N_s} f_0\left(J_0, \frac{V_s}{N_s}\right) \tag{3.13}$$

Here, V_s and I_s are the total voltage and current at the solar cell outlet. Analogously, dividing the total area of the electrolyzer electrodes into N_e series-connected modules, i.e., unit electrolyzers, yields:

$$I_e = \frac{S_e}{N_e} g_0\left(\frac{V_e}{N_e}\right) \tag{3.14}$$

Here, V_e and I_e are the total voltage and current at the electrolyzer.

During the operation of the "solar cell + electrolyzer" plant, Eq. (3.10) should hold. Whence it follows:

$$\frac{S_s}{N_s} f_0\left(J_0, \frac{V}{N_s}\right) = \frac{S_e}{N_e} g_0\left(\frac{V}{N_e}\right) \tag{3.15}$$

The solution of the optimization problem for the "solar cell + electrolyzer" plant consists in finding the N_s and N_e values that would ensure a maximum hydrogen-production rate in the plant, expressed as $P = I_e N_e$ (in electrical units), at certain areas[6] S_s and S_e and the irradiation intensity J_0.

In this plant, optimum performance is attained at a definite ratio of parameters N_s/N_e. To find these parameters, we transform Eq. (3.15) to:

$$\frac{S_s}{S_e} f_0(J_0, y/t) = t g_0(y) \tag{3.16}$$

where $t = N_s/N_e$ and $y = V/N_e$. The solution to Eq. (3.16) defines the function $y(t, J_0, S_s/S_e)$.

The maximum of $P = I_e N_e = S_e g_0(y)$ corresponds to the maximum of $y(t)$ because $g_0(y)$ is a monotonically increasing function of y. Therefore, the optimum condition for the plant performance is $y(t) = 0$. The variable $t = N_s/N_e$ at the integer values of N_s and N_e assumes a set of discrete values which at sufficiently large N_s and N_e can be regarded as quasi-continuous.

---

[6] At this stage of solving the optimization problem these areas are taken arbitrarily.

Constant radiation intensity. The above-obtained result makes is possible to develop, as the first stage of the solution of the optimization problem, an effective algorithm for the determination of optimum operating conditions. First, the value of maximum power generated by the cell and the value of y/t are found from the known current-voltage characteristic. Then, from Eq. (3.16) the value of t* corresponding to the optimum theoretical ratio N_s/N_e (in the general case, t* is a transcendental number), and the value of maximum theoretical output of the "solar cell + electrolyzer" plant are found. Finally, for a certain preset limiting value of $(N_s)_{max}$, the integer values of N_s^* and N_e^* giving the N_s/N_e ratio closest to t* are found and the plant output conforming to this ratio is calculated by using Eq. (3.15).

To solve this problem, a computer program in BASIC was developed [91] and calculations were performed. The program permits first of obtaining analytical approximations for the experimentally measured characteristics f_0 and g_0 in the form of the 6th order polynomials and, then, of computing the N_s^* and N_e^* values giving maximum P through numerical solution of Eq. (3.15) for various combinations of N_s and N_e. As mentioned earlier, the calculations are performed by pre-assigning the maximum admissible value of $(N_s)_{max}$ that limits the search range. This value enables one to efficiently realize the plant design (it is not convenient to divide the solar cell and electrolyzer into a very large number of modules). This particular computation (as well as other calculations in this section) was performed for the experimental plant [83] with $S_s = 950$ cm^2 and $S_e = 50$ cm^2. For the chosen value of $(N_s)_{max} = 15$, the optimum values of N_s and N_e were found to be equal to 9 and 2, respectively.

The computed function $P(N_s, N_e) = I_e N_e$ is shown in Fig. 49. It represents a "hangar roof"-type surface whose highest point lies above the straight line $N_s/N_e = t^*$. The point of intersection of the surface P and the normals to the (N_s, N_e) plane, drawn at the points conforming to integer values of N_s and N_e, gives the output of a real plant.

Varying radiation intensity. The second stage of the solution of the optimization problem resides in calculating the optimum design parameters at varying (e.g., during daylight) irradiation intensity at the solar cells surface. The areas S_s and S_e and the dependence $J_0(\tau)$, where τ is the time, are pre-set. The variation in solar

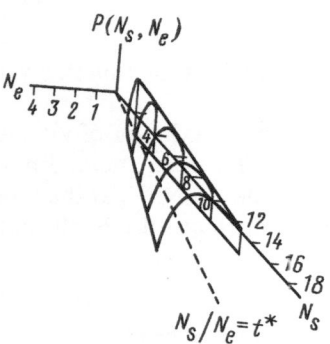

Fig. 49. Output P of plant (in arbitrary units) as a function of N_s and N_e [91]

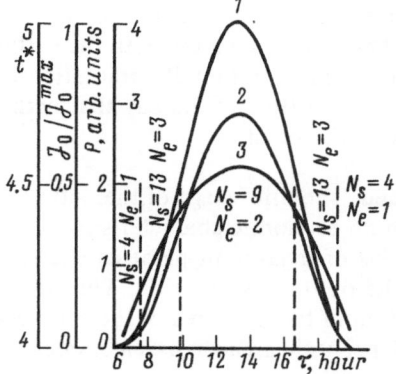

Fig. 50. Optimization of plant for varying light intensity
1 – dependence of relative irradiation intensity J_0/J_0^{max} on daytime (after curve 3 of Fig. 48); 2 – dependence of plant output P (in relative units) on τ (for gradually varying t); 3 – dependence of optimal ratio $t^* = N_s/N_e$ on τ
Dashed lines show time intervals with different optimal sets of N_s and N_e

radiation during the day in the absence of cloudiness depends on the latitude of the locality and the date. Now the task is to compute the dependences[7]:

$$t^*[J_0(\tau)] = [N_s(\tau)/N_e(\tau)]_{opt}$$

and

$$P(\tau) = (I_e N_e)_{opt}$$

The $t^*(\tau)$ curve shows how the automatic tracking servo system should vary the design parameters N_s and N_e during the day.[8] The computed dependences $t^*(\tau)$ and $P(\tau)$ and also the curve $J_0(\tau)$ used in the computation are shown in Fig. 50. In the calculations, $(N_s)_{max}$ was taken equal to 15. Curve 3 was used to find the instants when the system should be switched over to the new optimal parameters N_s and N_e according to the scheme:

$$\left(\frac{N_s}{N_e}\right)_{opt} = \frac{4}{1} \to \frac{13}{3} \to \frac{9}{2} \to \frac{13}{3} \to \frac{4}{1}$$

A variant of solving this optimization problem consists in calculating the optimal values of t^* and (for the preset $(N_s)_{max}$) of N_s and N_e in the absence of an automatic tracking system for switching over to new design parameters. The optimality criterion is the maximum daily integral hydrogen production of the plant:

$$\langle P(t) \rangle = \frac{1}{\tau_{max} - \tau_{min}} \int_{\tau_{min}}^{\tau_{max}} P[J_0(\tau), t] \, d\tau \tag{3.17}$$

---

[7] In the computations, which are of illustrative nature, for simplicity's sake, $f_0(J_0, v)$ is assumed to depend linearly on the irradiation intensity J_0. Taking into account the more complex dependences of f_0 on J_0 (including the indirect effect through solar cell heating) presents no fundamental difficulties and can be carried out with the use of analytical approximations of the tabulated $f_0(J_0, v)$ characteristics for several different irradiation levels.
[8] Do not confuse this system with the matching device discussed in the previous section; it does not allow the power generated by the solar cell to pass through it. Therefore, its operation does not involve appreciable energy losses.

For the plant having the characteristics listed in Ref. [83], maximum daily hydrogen production is attained at $N_s/N_e = 9/2$. This is in agreement with the earlier found solution of the optimization problem for the maximum daily irradiation intensity. It is not surprising because it is the brightest time of the day that makes the main contribution to the integral capacity of the plant.

Optimization by the cost of hydrogen produced. The third stage of the optimization problem involves the calculation of optimum economic characteristics of the "solar cell + electrolyzer" plant. The conditions of maximum plant capacity referred to the cost of the system was used as the optimality criterion. The cost of the system was calculated conditionally from that of the unit area of the solar cell u_s and the cost of the electrolyzer estimated per unit area of its electrodes u_e. Then:

$$F = \frac{P(S_s, S_e, J_0)}{u_s S_s + u_e S_e} \tag{3.18}$$

The plant capacity appearing in Eq. (3.18) is calculated as optimum capacity (in the sense that N_s and N_e are adjusted to the values of S_s and S_e) at a certain solar radiation intensity J_0. For the given values of u_s and u_e, the found dependence $F(S_s/S_e)$ has a maximum at a certain $(S_s/S_e)^*$. The optimum value of $(S_s/S_e)^*$ and, hence, of $F(S_s/S_e)^*$ depend on the current price of the solar cell and the electrolyzer. (The plant maintenance costs can be readily included in this price.)

The current prices depend on the type of electrolyzer or solar cell and vary rapidly with improvement in technology and changes in the market situation. At present, u_e/u_s can be taken approximately equal to 1. Indeed, as per the data of Ref. [93], 1 cm² of a silicon solar cell cost 5 US cents in 1983; the same is the specific price of electrolyzer with a solid polymer electrolyte. The price of an alkaline electrolyzer is somewhat higher, but of the same order. As an illustration, Fig. 51 shows the dependence of the optimal ratio of total areas of a solar array and electrolyzer on the ratio of specific prices u_e/u_s calculated for the unit described in Ref. [83]. Using this dependence, one can predict the design parameters of the plant depending on the market prices.

This approach is applicable for optimization not only by cost, but also by any other characteristic (in particular, weight) of the plant which can be taken to be approximately proportional to the areas of the solar cell and the electrolyzer electrodes.

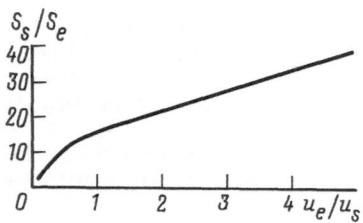

Fig. 51. Dependence of optimal ratio of areas of solar cell and electrolyzer on the ratio of specific prices of electrolyzer and solar cell [91]

3.5 Oxide Photoanodes for Photoelectrolysis of Water

3.5.1 Introduction. Main Problems

The photoelectrochemical cells intended for obtaining hydrogen at the cost of solar energy are mostly made, from the very beginning right up to the present, with oxide semiconductor photoanodes. In particular, the TiO_2 electrode served (and continues to serve) as a model electrode for studying the photoelectrolysis of water. The main advantage of such materials is that they, usually being higher oxides, do not degrade even under high anode potentials.

The disadvantage of most oxides is the very large width of the forbidden band: for TiO_2 about 3 eV, $SrTiO_3$ – 3.2 eV, $BaTiO_3$ – 3.3 eV, $KTaO_3$ – 3.5 eV. Therefore, these materials are sensitive only to ultraviolet light which is almost absent in the solar spectrum, especially near the Earth's surface. Hence, they by themselves are unsuitable for the conversion of solar energy (but not, for example, of more hard radiation, see Sect. 3.5.5).

Repeated attempts were made to select materials for photoanodes from semiconductor oxides having a much narrow forbidden band and, hence, sensitive to visible light. (It is the stability of oxides that makes them more attractive for the purpose.) However, these attempts have until now failed to overcome perhaps the main difficulty. Namely, in the oxide semiconductors in aqueous solutions, the top of the valence band, $E_{V,s}$, formed by 2p orbitals of oxygen is somehow "pinned" in the energy scale (probably, due to the interaction between the oxygen of water and the oxygen in the crystal lattice of the oxide). And it is located much below the electrochemical potential level of the H_2O/O_2 system (Fig. 52); this leads to unnecessarily large losses of energy during the energy conversion process (see p. 53). Therefore, on changing over to narrower bandgap oxides (WO_3, Fe_2O_3, and others), the location of $E_{V,s}$ remains practically unchanged and the conduction band bottom $E_{C,s}$ lowers. Simultaneously, the flat band potential of all oxides, with only one exception (see below), shifts towards more positive values according to the formula [95]:

$$\varphi_{fb}(NHE) = 2.94 - E_g(V) \tag{3.19}$$

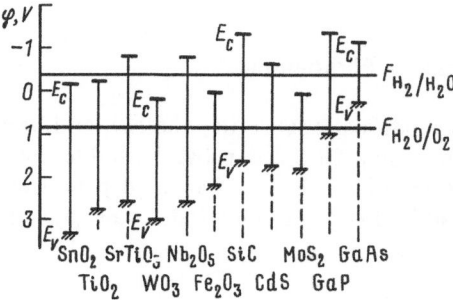

Fig. 52. Energy levels of band edges, E_C and E_V, of some semiconductors relative to electrochemical potential levels of water reduction and oxidation reactions (at pH 7) [24]

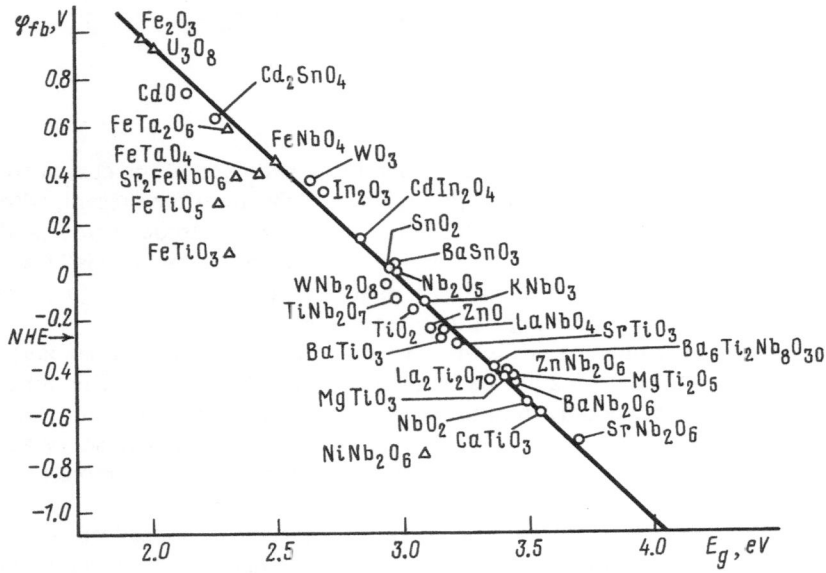

Fig. 53. Dependence of flat band potential of oxide semiconductors in aqueous electrolyte solutions on forbidden bandwidth
Δ – oxides with partially filled d-levels. Potentials are given against a saturated calomel electrode. NHE stands for normal hydrogen electrode potential

(Fig. 53) and very soon becomes more positive than the hydrogen-electrode reversible potential. Therefore, the energy of electrons in the conduction band proves to be insufficient for the evolution of hydrogen from water. As a result, the gain in sensitivity to visible light due to the decrease in E_g is overcompensated for by the loss caused by external voltage tapping off.

Methods have been proposed for sensitizing wide bandgap semiconductors to visible light by altering the non-equilibrium carrier's appearance mechanism, namely (a) by introducing certain impurities into the semiconductor (so as to replace the "intrinsic" absorption of light with the impurity absorption in the semiconductor), and (2) by adding dyes into the electrolyte (here, use is made of photoexcitation not of the semiconductor but of the dye molecule with subsequent transfer of an electron or a hole from the excited dye to the semiconductor). Both these methods are discussed in Sect. 3.5.3.

3.5.2 Most Important Oxide Semiconductor Photoanodes[9]

TiO$_2$. Titanium dioxide exists in crystal modifications: rutile, anatase, and brookite. Also, it is available as amorphous material. Of these, rutile (forbidden band width – 3 eV) exhibits the highest photosensitivity. But its flat band potential is slightly more positive than the reversible hydrogen electrode potential (see

---

[9] See also Sect. 9.2 in Ref. [1].

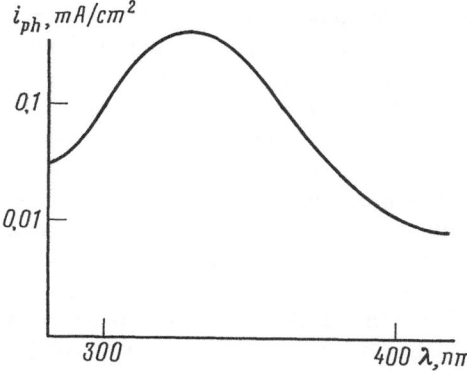

Fig. 54. Spectral dependence of anodic photocurrent of TiO_2 electrode obtained by oxidizing titanium foil in O_2 at 600 °C

Fig. 53), therefore spontaneous photoelectrolysis of water in the cell with a rutile photoanode does not take place. Photoelectrolysis occurs only on applying an external voltage (usually, about 0.5 V).[10] The forbidden bandwidth of anatase is slightly more (3.2 eV) and its flat band potential is by 0.1 V more negative than the reversible hydrogen electrode potential, which makes it possible to photodecompose water even without the application of an external voltage. This is particularly important not for macroscopic photoelectrochemical cells but for semiconductor suspensions (see Sect. 5.3) where it is not possible to apply an external voltage.

The weak point of TiO_2 electrodes is, as mentioned earlier, their very wide forbidden band (the threshold wavelength for the absorption of light and the appearance of photocurrent is about 400 nm). Besides, the indirect nature of interband electron transitions in TiO_2 is responsible for a very gradual growth in the light absorption coefficient α with quantum energy. Therefore, α attains its maximum value at a light wavelength of 340 nm [96] (cf. Fig. 54). Only short-wave light is absorbed at a depth of the order of space-charge-layer thickness (about 10^{-6} cm); more long-wave light is absorbed at a greater depth from where the generated holes cannot reach the electrode surface (in polycrystalline TiO_2 the hole diffusion length is of the order of 10^{-6} cm [97]) and recombine without making contribution to the photocurrent.

Monocrystalline samples exhibit best photoelectrochemical characteristics: maximum quantum yield of photocurrent (at $\lambda = 330$ nm) approaches 1. And polycrystalline samples with good photoelectrochemical activity can be prepared often by simple methods.

The main methods of preparing polycrystalline TiO_2 samples involve[11]: thermal and anodic oxidation of metallic titanium; chemical vapor deposition of

---

[10] Starting from the first report on photoelectrolysis of water [48] a difference in the pH of anolyte (for example, 1 M NaOH) and catholyte (0.5 M H_2SO_4) of the cell was created and the electromotive force of the concentration cell so formed was used instead of applying an auxiliary battery voltage. This emf must be substituted for φ_{ext} in Eq. (3.3) to compute the photoassisted electrolysis efficiency.

[11] Most of the enumerated methods are suitable also for obtaining Fe_2O_3, WO_3, and other oxide electrodes.

TiO_2, including the application of RF power; cathode sputtering of TiO_2 (or metallic titanium in an oxygen-containing atmosphere); pyrolysis of titanium salts or organotitanium compounds pre-deposited on the substrate; pressing and sintering of TiO_2 powder (in the form of ceramic tablets); smearing the active mass (TiO_2 powder mixed with the binder) on the substrate. Thin-film as well as ceramic [98] samples are best photoelectrodes (quantum yield of photocurrent is about 0.7). The former are obtained by thermal oxidation of titanium (in air or O_2 at 500–700 °C) or by anodic oxidation of titanium with subsequent heating in vacuum [96] and by plasma deposition [99]. As prepared by the enumerated methods, TiO_2 exhibits the properties of an insulator, as a rule. For rendering sufficient conductivity to electrodes, donors are introduced into TiO_2: oxygen vacancies or Ti^{3+} ions (usually by heating in vacuum or hydrogen), protons (by cathodic polarization in aqueous solutions). These simple methods have, however, a disadvantage: both oxygen vacancies and protons in particular are highly mobile in TiO_2 and, when the photocurrent is passed, they gradually migrate towards the TiO_2/solution interface where the vacancies get filled with oxygen evolving at the electrode, and the protons get oxidized. As a result, the photocurrent gradually decreases. More stable are electrodes with less mobile donors (for example, Nb) which are introduced into TiO_2 in manufacturing electrodes.

The efficiency of water photoelectrolysis cells with a titanium dioxide photoanode under sunlight does not exceed 1 % (see, for example, Ref. [100]).

$SrTiO_3$, $BaTiO_3$, $KTaO_3$. In these wide bandgap semiconductors the conduction band bottom lies above the water reduction electrochemical potential level in aqueous solutions, F_{H_2/H_2O} (Fig. 52). That is why they are capable of decomposing water without the application of an external voltage. Monocrystalline $SrTiO_3$ is noted for maximum activity in water photoelectrolysis: for monochromatic light (quantum energy 3.8 eV) the efficiency equals 20 %. For sunlight, the efficiency is of course very small.

In the case of $SrTiO_3$ as well as TiO_2 and other wide bandgap semiconductors, the difference between F_{H_2O/O_2} and $E_{v,s}$ is very large (about 1.5 eV). Therefore, direct electron transition between the semiconductor valence band and the filled levels in the solution seems to be less probable. The transitions are assumed to take place via a certain surface level (cf. Fig. 32) located in the semiconductor forbidden band.

WO_3, Fe_2O_3, In_2O_3. These oxides with narrower forbidden bands (2.7, 2.2, and 2.8 eV, respectively[12]) are sensitive to visible light. But for the aforementioned reasons (the flat band potential is more positive than the reversible hydrogen electrode potential) they are suitable for splitting water only under photoassisted electrolysis conditions. The efficiency of this process under sunlight is less than 1 %.

At the indium oxide electrode (obtained by thermal oxidation of indium metal) the quantum yield of photocurrent amounts to 0.9 (for a light wavelength of 310 nm) [101]; at the tungsten oxide electrode, it reaches 0.6 (350 nm). The stability of Fe_2O_3 and WO_3 is somewhat worse than that of TiO_2 or $SrTiO_3$.

---

[12] The value of E_g somewhat varies depending on the electrode material manufacturing technique.

3.5.3 Sensitization of Wide Bandgap Oxides

Sensitization by dyes in solution. Behind this method of photosensitization are the photoelectrochemical reactions on a semiconductor electrode, involving the participation of excited dye molecules (ions) from the solution. The photoelectrochemistry of excited states is a wide field of physical chemistry (see, for example, Ref. [1] Ch. 5 and Ref. [102]). Here, we shall consider it only to the extent at which it may help in solving the solar energy conversion problem.

Suppose that the electrolyte solution in the electrochemical cell contains a substance which absorbs visible light. (Hence, such substances are dark-colored.) Having absorbed a light quantum, the dye particles in the solution get excited. As the lifetime of the excited state is more than the time required to reach equilibrium with the surroundings (and, especially, the time for electron transfer between the particle and the electrode), one can treat the excited particles in the same way (cf. Sect. 1.3) as ordinary chemical reagents, i.e., they can be assigned, in analogy with E_{ox}^0 and E_{red}^0, most probable energy levels for reduced and oxidized forms, E_{ox*}^0 and E_{red*}^0 (with corresponding $W(E)$ distributions), and the electrochemical potential level F_{redox*} (Fig. 55). The difference between the corresponding energy levels of the ground and excited states is the excitation energy ΔE.

The reversible electrode potentials of the oxidation and reduction reactions of excited dye particles S*, related to the electrochemical potential levels F_{S*/S^+} or F_{S*/S^-} by Eqs. (1.13), differ from the reversible potentials of the same reactions with the participation of unexcited molecules, S, by the value of the excitation energy [102]:

$$S^* \rightarrow S^+ + e^-, \qquad \varphi_{S*/S^+}^0 = \varphi_{S/S^+}^0 - \Delta E/e \qquad (3.20\,a)$$

(cf. Fig. 55);

$$S^* + e^- \rightarrow S^-, \qquad \varphi_{S*/S^-}^0 = \varphi_{S/S^-}^0 + \Delta E/e \qquad (3.20\,b)$$

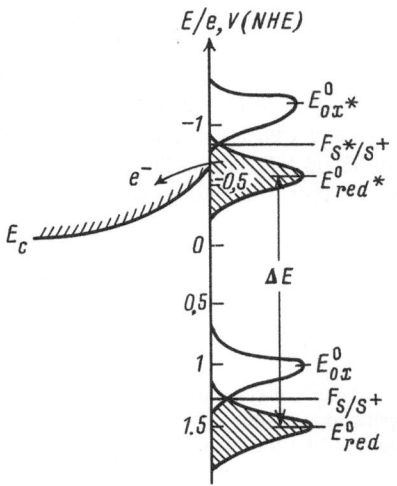

Fig. 55. Schematic of the process of sensitizing TiO$_2$ semiconductor with Ru(bpy)$_3^{2+}$ dye in aqueous electrolyte solution (pH 10)
The valence band edge of TiO$_2$ is not shown (for the chosen energy scale it did not find room in the figure)
S and S$^+$ stand for the reduced and oxidized form of the dye

(For further discussion, the most interesting case is $\Delta E < E_g$; here, illumination excites just the solution, leaving the semiconductor electrode unexcited.)

If the distribution of levels of the excited dye are overlapped by the allowed energy bands of the semiconductor electrode and those of the unexcited dye are not overlapped, then the transfer of electrons between the electrode and solution, i.e., the occurrence of the electrochemical reaction, is possible only under illumination. For example, in the case shown in Fig. 55 the distribution of electron-occupied levels of the reduced form of the dye, bipyridyl complex of ruthenium (II) $(Ru(bpy)_3^{2+})$ which is an electron donor, in the ground state lies opposite to the forbidden band of the semiconductor (TiO_2). Therefore, $Ru(bpy)_3^{2+}$ is not oxidized at this electrode in the dark. Upon illumination, $Ru(bpy)_3^{2+}$ gets excited (i.e., the electron rises to a higher energy level) and the distribution of occupied levels of the excited state is found to be opposite to the conduction band of TiO_2. From the excited $Ru(bpy)_3^{2+*}$ ion the electron goes into TiO_2 (and is then carried over from the surface into the electrode bulk owing to the existence of an electric field in the space charge region, which makes the interphase electron transfer irreversible). A photocurrent now flows in the cell; what is more, this occurs in the more long-wave light region compared to the TiO_2 intrinsic absorption region where light is absorbed by the semiconductor (Fig. 56).

If the solution contains, besides the dye, a substance (known as supersensitizer) capable of reducing the oxidized form $(Ru(bpy)_3^{3+})$ produced at the electrode, then the dye is recovered and re-enters into the photoelectrochemical reaction. The entire process amounts to oxidizing the supersensitizer; in this case, $Ru(bpy)_3^{2+}$ acts as a photocatalyst because the supersensitizer as such cannot get oxidized at the semiconductor photoanode.

The most frequently used supersensitizers are I^-, Br^-, hydroquinone, and other reductants. Of course, for effective sensitization of a semiconductor in the process, precisely, of water photoelectrolysis, it is necessary that water itself should act as supersensitizer. In fact, certain dyes (e.g., the bipyridyl complex of ruthenium) in the oxidized state are capable of liberating oxygen from water; however, this process proceeds slowly and has not yet been used for practical purposes.

Another difficulty is the very short lifetime of the excited state of the dye. Therefore, only the dye which is present in the immediate vicinity of the electrode surface, and even better – the one adsorbed on the electrode – proves to be electroactive. But the absorption of light in the adsorbed dye layer, due to its small thickness, is poor even when the extinction coefficient is large. That is why the quantum yield of sensitized photocurrent does not exceed a few percent (referred

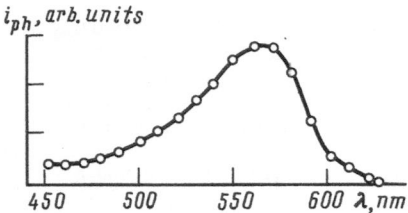

Fig. 56. Photocurrent spectrum of the TiO_2 electrode sensitized to visible light using $Ru(bpy)_3^{2+}$ [103] (Reprinted by the permission of the publisher, The Electrochemical Society, Inc.)

to the incident light), and one does not expect a large light-energy conversion efficiency. Relatively thick layers of dyes capable of absorbing the entire incident light cannot be used, because, being dielectrics, the dyes in a thick layer electrically insulate the electrode from the solution.[13] That is the reason why sensitization of semiconductors to long-wave light is until now effectively employed in those cases where complete utilization of light energy is not necessary, e.g., for recording information, in photochromic devices, in photography, etc., but not for light energy conversion. Possibly, special designs of photoelectrodes in which light multiply passes through the adsorbed monolayers of the dye (e.g., through a porous semiconductor body impregnated with the dye solution or a set of plane parallel thin-layer electrodes with the adsorbed dye, see Ref. [104]) shall prove more effective in making use of a solution's photoexcitation for light energy conversion.

The best sensitizers for semiconductor electrodes are well-adsorbing dyes (for example, cyanine dyes). In a concise form, the results of investigations into dye sensitization of semiconductor electrodes (not only oxides) are given in Ref. [105] and Ref. [2] Chap. 11; much earlier works are reviewed in Ref. [106].

Note that, as in photogeneration of electron-hole pairs due to "intrinsic" absorption of light in the semiconductor (see Sect. 2.4), there is some optimum concentration of donors (acceptors) in the semiconductor that ensures most effective separation of charges. Thus, for the SnO_2 and ZnO electrodes sensitized, respectively, by Rhodamine B and Bengal rose, the quantum yield of sensitized photocurrent is maximum at $N_D = 3 \times 10^{20}$ cm^{-3}. At smaller concentrations of donors, the space charge layer is too wide, the field therein is small and part of the electrons injected in the semiconductor recombine; at larger values of N_D the space charge layer is very narrow and the electrons from the semiconductor can tunnel through this layer back to the solution [107].

Sensitization by impurities in the semiconductor is based on the effect of current-carriers photogenerated by photoionization of impurity atoms but not by interband transitions. This mechanism is briefly discussed in Sect. 1.2. As already mentioned, the free carriers' generation efficiency is not great because of poor absorption of light by the impurity due to its small concentration (usually up to several atomic percent). In the literature one can find numerous works on sensitization of TiO_2, $SrTiO_3$, and other wide bandgap oxides with admixed Al, Cr, Mn, and a large number of other metals (a brief review on such works is available in Ref. [105]). Though the results of these studies are often contradictory (which in many respects is due to poorly controlled sample preparation techniques), a main conclusion can yet be made: the introduction of an impurity often widens the spectral sensitivity range of the matrix oxide towards long waves. However, (1) the quantum yield of sensitized photocurrent is small and (2) there is a pronounced decrease in the quantum yield of photocurrent in the "intrinsic" absorption region (i.e., for the UV light) where it was quite significant prior to the introduction of impurities. This situation, typical for diverse combinations of semiconductor/im-

---

[13] Here, we shall not dwell on the use of thick (phase) layers of pigments like chlorophyll, phthalocyanine, and others, after imparting them electric conduction, as individual photoelectrodes.

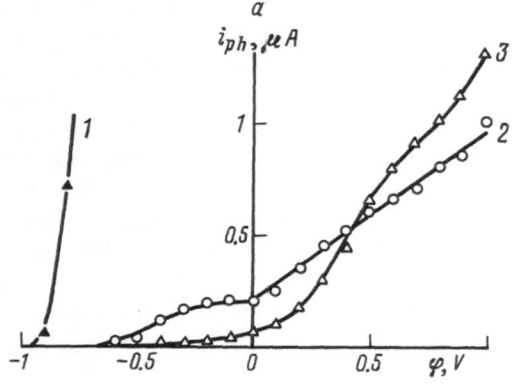

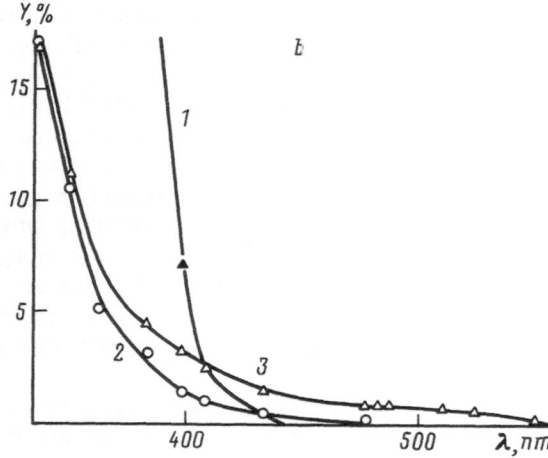

Fig. 57. Dependence of photocurrent of Cr-doped TiO_2 electrode on potential (a) and of
quantum yield of photocurrent
on light wavelength (b) in $1 N$
NaOH [98] Ceramic electrodes
were prepared by pressing and
baking the mixture of powder
oxides. The Cr content (in %):
1 – 0; 2 – 1; 3 – 5

purity, is shown in Fig. 57, taking doping of TiO_2 with chromium as an example
[98].

The decrease in photocurrent in the intrinsic absorption region is generally associated with the fact that doping causes the recombination rate to increase.

3.5.4 Photoanodes – Mixed Oxides with Two Cation Sublattices

In the foregoing it has been mentioned that the unfavorable relationship between
the forbidden bandwidth and the flat band potential (see Eq. (3.19)) does not permit oxide semiconductors to be used directly as photoanodes in the photoelectrolysis of water. The approach briefly discussed below [50, 108–110] makes it possible, in principle, to develop photoanodes which, by retaining the advantages of
wide bandgap oxide semiconductors (negative value of flat band potential, chemical stability) would be free of their main disadvantages: insensibility to visible
light, very low location of the valence band top relative to the F_{H_2O/O_2} level. It in-

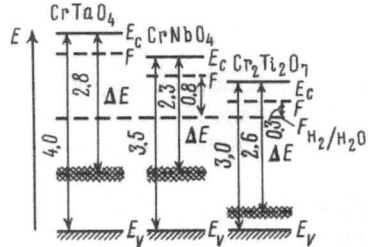

Fig. 58. Approximate energy diagram of some mixed oxides. Double hatch – upper edge of valence d-band. The difference of energies is given in electron volts. The location of valence "oxygen" 2p-band top is conditionally taken the same in all oxides

volves synthesis of mixed oxides in which both the allowed energy bands – valence and conduction bands – can be created by the orbitals either of the same or two different cations.

The example of an oxide in which both the allowed energy bands are formed by the same cation (Rh) is $LuRhO_3$ – a semiconductor with a forbidden bandwidth $E_g = 2.2 \text{ eV}$. (It exhibits p-type conductivity and may therefore be used not as a photoanode but as a photocathode.) However, the band edges of $LuRhO_3$ are unfavorably located relative to the electrochemical potential levels of the water-decomposition reactions: the valence band top lies above the F_{H_2O/O_2} level, therefore the photogenerated holes in $LuRhO_3$ are unable to liberate oxygen from water without the application of an external voltage. That is why the photoelectrolysis efficiency is small [111].

The formation of allowed bands by two different cations may be observed, e.g., in mixed oxides composed of TiO_2, Nb_2O_5, and Ta_2O_5, on the one hand, and of Cr_2O_3, on the other hand. The orbitals of the first cation (Ti, Nb, Ta) form a conduction band which is located relatively high, whereas the d-orbitals of the second cation (Cr) form an additional valence band; the latter band lies in the forbidden band of the matrix oxide by 0.5–1.5 eV higher than the "oxygen" 2p-band (Fig. 58). The photoelectrolysis of water occurs owing to the excitation of electrons of this valence band and their transition to the conduction band. The study of the photoelectrochemical behavior of a number of materials ($Cr_2Ti_2O_7$, $CrNbO_4$, and others) has revealed that the threshold energy ΔE at which photocurrent appears is in fact less than that for the TiO_2, Nb_2O_5, or Ta_2O_5 matrix. Therefore, the obtained substances are sensitive to visible light. At the same time, the flat band potentials – unlike the narrow band oxides – are more negative than the reversible hydrogen potential (Table 3.3, cf. Fig. 23). The relationship between ΔE and φ_{fb}, unlike the relationship between E_g and φ_{fb}, is no more expressed by Eq. (3.19), see Fig. 53. This provides a basic way of carrying out photoelectrolysis of water under

Table 3.3. Characteristics of mixed-oxide photoanodes [109]

	$CrNbO_4$	$CrTaO_4$	$Cr_2Ti_2O_7$
ΔE, eV	2.3	2.8	2.6
φ_{fb}, V (against NHE)	−0.8	−1.3	−0.35

the action of visible light but without the application of an external voltage. Nevertheless, no success has so far been made in attaining a sufficiently high quantum yield of photocurrent, the reason being, perhaps, a large recombination rate as a consequence of the imperfect structure of the studied ceramic electrodes.[14] Upgrading the manufacturing technique of photoanodes, primarily the change-over to large-crystalline (the more so, monocrystalline) materials may make them more suitable for practical use.

The authors of Ref. [112] have reported a high efficiency of $LaCrO_3$ photoanodes, but this information needs further verification.

3.5.5 Conversion of Ionizing Radiation Energy

As upon illumination, electric current appears in the cell with a semiconductor electrode under the action of ionizing radiation (accelerated electrons, γ-radiation, neutron radiation, X-rays). In the first approximation, the nature of this phenomenon is the same as that of the photoeffect considered above: the radiation-electrochemical current is caused by the holes and electrons generated upon absorption of radiation energy.

It must be borne in mind that the energy, e.g., of γ-quanta amounts to 10^5-10^6 eV, i.e., it is by many orders more than the quantum energy of electromagnetic radiation in the visible and ultraviolet ranges of the spectrum with which one is concerned in semiconductor photoelectrochemistry, and the forbidden bandwidth of semiconductors. Therefore, one γ-quantum gives rise not to one but many electron-hole pairs in the semiconductor. Moreover, here, compared to the earlier considered photoelectrochemical reactions, the complicating features are: radiolysis of the solution and radiation damage of the semiconductor; their consequences are the often observed relaxation phenomena, e.g., gradual change in current density and others. But some wide bandgap oxide semiconductors (for example, $SrTiO_3$, TiO_2) have proved to be good radiation-resistant electrode materials for making anodes for radiation-electrochemical cells [113].

By analogy with the photoelectrochemical conversion of light energy, the appearance of a radiation-electrochemical current can be considered as the basis of the ionizing radiation energy conversion method [114]. Indeed, the values of conversion efficiency attained so far are insignificant: 10^{-4}-10^{-2} %; these values have been obtained for water radiation-assisted electrolysis in unoptimized cells.

The cause of such low values of efficiency is that, owing to large penetrability of the ionizing radiation, only a minute fraction of incident radiation energy is absorbed in the space charge layer (thickness 10^{-6} to 10^{-5} cm) in the semiconductor where separation of irradiation generated charges takes place. The radiation-electrochemical energy conversion efficiency can be increased, for example, by replacing the compact semiconductor electrode with a porous electrode. In so doing, the fraction of semiconductor volume occupied by the space charge layers at the sur-

---

[14] In some cases, the cause of low photocurrent may, apparently, be more serious: the valence band formed by d-orbitals is found to be narrow, and, therefore, the holes therein have low mobility.

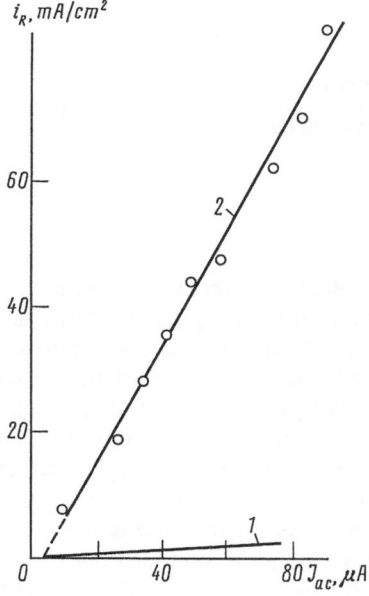

Fig. 59. Dependence of radiation electrochemical current (i_R) of water oxidation at smooth (1) and porous (2) TiO$_2$ anode on the intensity of flux of accelerated electrons with energy 4 MeV
Along x-axis – electron accelerator current (J_{ac}) proportional to the density of energy incident on the cell

face of pores increases. That is why in a porous semiconductor electrode the entire volume is more effectively used during the radiation electrochemical process; as a result, the current is much higher than on the smooth electrode. As an illustration, Fig. 59 shows how the current increases approximately by two orders (i.e., the efficiency grows from 10^{-4} to 10^{-2}%) on going from the smooth TiO$_2$ electrode to the porous electrode [113]. There is every reason to believe that the radiation electrochemical water decomposition efficiency can be further enhanced, e.g., by increasing the semiconductor/solution mass ratio and also by improving the composition and structure of semiconductor electrodes and the design of cells.

Chapter 4
Solar Energy Conversion into Chemical Energy.
Cells for Photoelectrolysis of Inorganic
(other than Water) and Organic Substances

Attempts have been made to apply numerous photoelectrolysis reactions, as well as photoelectrolysis of water, for obtaining hydrogen from cheap and readily available reagents (sometimes even from waste products). These are the decomposition (to be more exact, dehydrogenation) reactions of hydrogen sulfide, hydrogen halide acids, and of some organic compounds. In other processes, fuel or raw material different from hydrogen is obtained: saturated hydrocarbons (in the so-called "photo-Kolbe reaction"), methyl alcohol (in the photoreduction of CO_2). Also, we shall mention the photoreactions of fixing molecular nitrogen and oxidizing certain substances that pollute water (for instance, SO_2). A distinguishing feature of most of these reactions is that their Gibbs' energy change is smaller than for the water decomposition reaction. For example, in decomposing H_2S, HI, and HBr into elements, the change of standard Gibbs' energy ΔG amounts, respectively, to 0.14, 0.54, and 1.07 eV. Thanks to smaller ΔG, these substances are decomposed by avoiding the difficulties typical of water photoelectrolysis under sunlight.

4.1 Photodecomposition of Hydrogen Halide Acids and other Inorganic Compounds

The HBr and HI photoelectrolysis products are hydrogen and a halide.[1] For practical purposes, the photoelectrolysis of HBr is of prime interest, because bromine and hydrogen can be readily stored and used afterwards for obtaining electrical energy with the aid of a bromine-hydrogen fuel cell.

In photoelectrolysis cells use can be made of both photoanodes (n-type $MoSe_2$, WSe_2, and Si) and photocathodes (p-type InP and Si). The indium phosphide surface is made catalytically active by depositing, as described above, the islets of a noble metal (Pt, Rh, Ru [115]) on it. Silicon electrodes are protected against photocorrosion, e.g., by covering with a platinum silicide film having "metallic" conductivity [116] or with a mixed oxide of ruthenium and titanium [117] (for detailed information on protective layers, see Sect. 7.1).

---

[1] For the HCl decomposition reaction the standard Gibbs' energy is large (1.36 eV). Therefore, this reaction, just as the decomposition of water, is carried out under the photoassisted electrolysis regime (see Sect. 3.2).

The quantitative characteristic of a photoelectrochemical cell for decomposition of HI and HBr can be readily obtained from Fig. 60, showing the current-potential curves for the H_2 and halide evolution reactions on the semiconductor electrodes (n-MoSe$_2$ and p-InP) under illumination, as well as on metal (platinum) electrodes in the dark. For the photoelectrolysis reaction to occur spontaneously (without the application of external voltage) the current-voltage characteristics of the anode and cathode should overlap each other in some range of potentials; in this case, the current and electrode potential can be easily found from the condition $i_a = -i_c$. This is what happens in the case of HI (Fig. 60a): photoelectrolysis proceeds spontaneously in the cell with only one photoelectrode (i.e., with InP-Pt or MoSe$_2$-Pt pairs) and, more effectively, with two photoelectrodes. But in the case of HBr, due to a more positive potential of bromine evolution (or, the same, due to a higher ΔG of the reaction), spontaneous photoelectrolysis is possible only in the cell with two photoelectrodes.

Photodecomposition of HBr is feasible also in a cell with only one photoelectrode, provided this electrode operates under the two-quantum electrolysis regime, for example, if it has an inner p-n-junction. On a n-type Si electrode, in which a p-n-junction has been created close to the electrolytic interface by the boron-ion implantation technique, spontaneous photooxidation of Br$^-$ proceeds effectively [118].

Both the enumerated principles – two photoelectrodes in one cell and two-

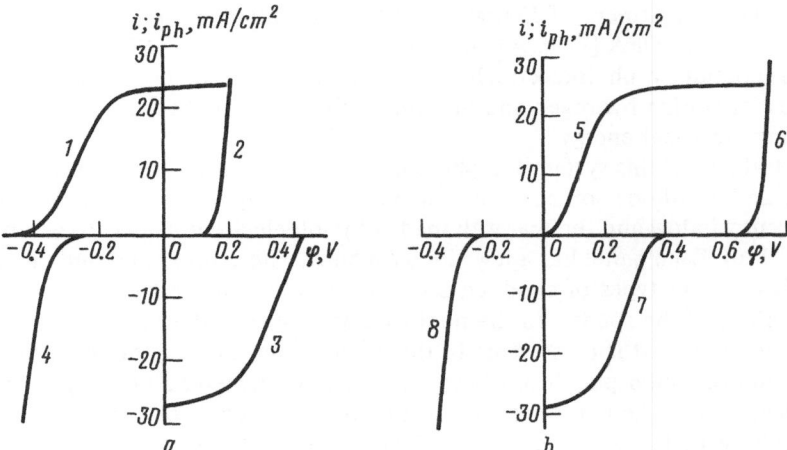

Fig. 60. Current-voltage characteristics of electrodes in cells for photoelectrolysis of HI (a) and HBr (b) [115]
1 - photooxidation of I$^-$ on n-MoSe$_2$; 2 - dark oxidation of I$^-$ on Pt; 3 - photoevolution of H$_2$ on p-InP(Rh); 4 - dark evolution of H$_2$ on Pt; 5 - photooxidation of Br$^-$ on n-MoSe$_2$; 6 - dark oxidation of Br$^-$ on Pt; 7 - photoevolution of H$_2$ on p-InP(Rh); 8 - dark evolution of H$_2$ on Pt
Solutions: $2\,M\,HI + 10^{-2}\,M\,I_3^- + 2\,M\,NaClO_4$ (a); $2\,M\,HBr + 10^{-2}\,M\,Br_3^- + 2\,M\,NaClO_4$ (b); anode and cathode compartments are separated by a diaphragm; halogen lamp is used for illumination
i and i_{ph} stand respectively for dark current and photocurrent (Reprinted by the permission of the publisher, The Electrochemical Society, Inc.)

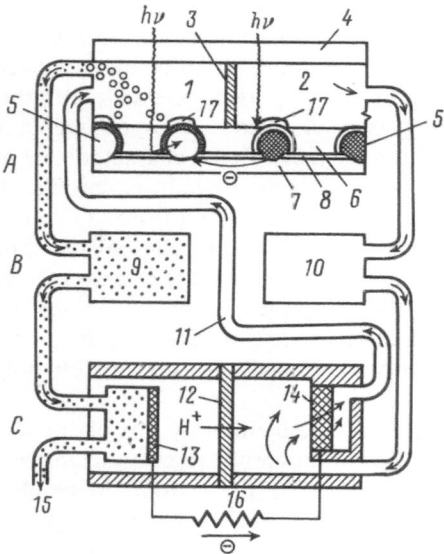

Fig.61. Scheme of TISES unit for photoelectrochemical decomposition of HBr (reproduced from Ref. [119])
A – cell for photoelectrolysis; B – H_2 and Br_2 storage devices; C – fuel cell
1 – cathode half-cell; 2 – anode half-cell; 3 – diaphragm; 4 – glass window; 5 – silicon spherical electrode (dark regions are of n-type; white regions are of p-type); 6 – glass matrix; 7 – current conducting bar; 8 – reflector; 9 – hydrogen accumulator; 10 – storage tank for bromine-rich electrolyte, and heat-exchanger; 11 – recovery of bromine-free electrolyte; 12 – cation-exchange membrane; 13 – gas diffusion hydrogen electrode; 14 – bromine electrode; 15 – excess gas outlet; 16 – external load; 17 – protective metal film on silicon

quantum photoelectrolysis regime – are integrated[2] into the Texas Instruments Solar Energy System (TISES) for conversion of solar energy into electrical energy (with its storage and subsequent liberation). A sum of 14 million dollars was allocated by the US Department of Energy for the development of these systems by Texas Instruments of USA [119]. A schematic illustration of the TISES is shown in Fig. 61. It contains a photoelectrochemical cell for the photodecomposition of HBr, devices for storing hydrogen and bromine, and a bromine-hydrogen fuel cell for generating electrical energy.

The central part of this system is a photoelectrochemical cell proper consisting of an anode and a cathode compartment separated by a diaphragm. The front wall is a transparent window and the rear wall made of photoelectrode-panels has a unique design [120]. Each panel has a few silicon spheres whose inner part and outer shell exhibit different types of conductivity. Therefore, a p-n-junction is formed under the surface of the sphere. In the photocathode spheres, the inner part is of p-type and the outer part is of n^+-type. In the photoanode spheres, the inner part is of n-type and the outer part is of p^+-type. Spheres of diameter 380–400 μm are "cast" like lead shots. The authors of this technology, the details of which are not disclosed, believe that it will enable very cheap photoelectrodes to be obtained. The spheres are pressed into a glass matrix such that their center-to-center distance exceeds 2.2 times the diameter. The matrix has a reflector on its rear side, therefore the light which passes through the electrolyte and is incident on the rear wall is reflected again to the surface of the sphere. The rear part of the spheres is grounded off and is pressed against the conducting (tantalum) bar so that the latter contacts with the inner part of each sphere but not with its outer shell. As a result, all the spheres are electrically connected to each other in such a manner that

---

[2] Compare the water photoelectrolysis cell described in Sect. 3.3.

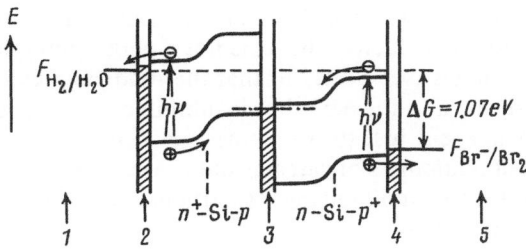

Fig. 62. Energy diagram of TISES unit when illuminated (reproduced from Ref. [119])
1 - electrolyte solution (catholyte); 2 - protective film of Pt-Ir; 3 - metal of conducting bar; 4 - protective film of iridium/iridium oxide; 5 - electrolyte solution (anolyte)

the cell is under short-circuit conditions. The outer surface of silicon spheres (in contact with 8.8 M HBr solution) is covered with a transparent protective-catalytic layer 10–40 nm in thickness; the cathode is covered with a platinum-iridium (in the ratio of 7:3) alloy and the anode, with iridium/iridium oxide film. The panel area is 38–40 cm^2.

According to the energy diagram of the cell, shown in Fig. 62, hydrogen evolves on the photocathode upon illumination and bromide is oxidized to bromine on the photoanode. The bromine-rich solution is drained from the anode half-cell into a special storage tank equipped with a heat exchanger, and gaseous hydrogen from the cathode half-cell, into an accumulator where it is stored in the form of a solid hydride of the CaNi$_5$ alloy. The hydride upon heating liberates hydrogen. To this end, use is made (with the aid of the heat exchanger) of solar heat concentrated in the bromine solution.

For obtaining electrical energy at the unit outlet, hydrogen and bromine (in the form of a solution) are supplied to the fuel cell; the worked-out electrolyte (depleted of bromine) is returned to the cathode half-cell.

At a radiation power density of 100 mW/cm^2 a photovoltage of 0.55 V is developed in every p-n-junction (for the cell as a whole this value equals 1.1 V); the current density of electrolysis amounts to 26–27 mA/cm^2. The overvoltage on electrodes is small: even at a current density of 150 mA/cm^2 it does not exceed 95 mV at the cathode and 65 mV at the anode [121]. A photoelectrolysis efficiency of 8.6 % has already been attained [120]. As reported by the authors of Ref. [119], in the final variant the "solar energy-to-electrical energy at the outlet" efficiency would amount to 10 %.

The photoelectrochemical characteristics of a TISES as a whole satisfactorily lend themselves to mathematical simulation. For this purpose, use is made of semiconductor properties of the electrodes, rate constants of electrochemical stages, values of resistances, etc. The studies made into corrosion of protective coatings make it possible to predict 20 years service life for panels [122, 123].

The investigations into other photoelectrochemical reactions involving the participation of inorganic compounds have not yet exceeded the laboratory stage. Among these reactions are: reduction of SO$_2$ on a p-InP electrode [124], dehydrogenation of H$_2$S on a CdSe electrode [68] (this reaction is discussed in greater detail in Sect. 5.4 as to the application of semiconductor suspensions) and fixation of molecular nitrogen. The latter reaction proceeds in a cell with a p-type GaP photocathode, an aluminium anode and a solution of AlCl$_3$ and titanium tetraisopropoxide in a nonaqueous solvent – glym (1,2-dimethoxy ethane). When the photocathode is illuminated, nitrogen is reduced to ammonia and the aluminium anode is dissolved. This reaction, unlike those considered above, is not a photosynthetic but a photocatalytic reaction. This implies (see Sect. 2.3) that thermodynamically the reaction can occur spontaneously even in the dark, but is inhibited due to kinetic reasons. Here the light energy is spent just to overcome this kinetic barrier [125, 126]. The reports about successful attempts of carrying out the same reaction as actual photosynthesis, i.e., without using a dissolving ("sacrificial") aluminium anode are perhaps erroneous (see Ref. [127]).

4.2 Photoelectrochemical Reactions of Organic Compounds

The photoelectrochemical oxidation or reduction reactions on semiconductor electrodes, with the participation of organic substances, can be utilized to synthesize useful products as well as to destroy harmful substances, say, in purifying waste waters. By now tens of such reactions have been studied to various extents, mainly, not in the photoelectrochemical cells with compact electrodes but in semiconductor suspensions. Some of these reactions are listed in Sect. 5.4. Here we shall mention some of their common pecularities.

The reactions of organic compounds both on semiconductor photoelectrodes and metal electrodes generally occur in several stages and yield a number of intermediate products. A consequence of this for photoelectrochemistry is the so-called "current doubling". This manifests itself in that the photocurrent measured in the oxidation or reduction reaction of any substance involving the participation of photogenerated minority carriers exceeds approximately twice the current of these carriers, which reaches the semiconductor surface. Most probably, this is because the minority carriers take part only in one (usually the first) stage of the reaction. The subsequent stages proceed with the involvement of majority carriers. Thus, during photooxidation of formaldehyde on a CdSe electrode, the reaction proceeds as per the scheme:

$$CH_2O + h^+ + OH^- \; \rightarrow \; HCOOH + H^{\bullet}$$
$$H^{\bullet} \qquad\qquad\quad \rightarrow \; H^+ + e^- \qquad\qquad (4.1)$$

Here, h^+ is a hole in the valence band; e^- is an electron in the conduction band of CdSe. Atomic hydrogen ($H^{\bullet}$) is the intermediate product. Would this scheme have been adequate for a real process, then the photocurrent would, in fact, have been twice as much as the hole current. Generally, the intermediate product is partially consumed in the course of side reactions with other components of the solution, during recombination, disproportionation, etc., and the photocurrent "multiplication factor" is found to be slightly less than 2.

The other distinguishing feature is due to the competition of different pathways of photoelectrochemical reactions. If the solution contains several reagents, then the intermediate products of their transformation can react with each other, yielding new products. For example, when the titanium dioxide photoanode in the sodium acetate and iodide solution is illuminated, parallel processes of photooxidation of CH_3COO^- and I^- ions take place and the formed $CH_3^{\bullet}$ and $I^{\bullet}$ radicals undergo cross coupling [128].

For years the photoreduction of CO_2 on p-type semiconductor photocathodes, as a process analogous to natural photosynthesis, has attracted the steady attention of researchers. Formic acid with an admixture of methanol and formaldehyde is known to be the main CO_2 reduction product on metal electrodes. The same happens in a photoelectrochemical cell with a p-GaP photocathode [129]. The reaction proceeds with some "underpotential" (compared to usual "dark" electrolysis); in this reaction the energy is stored with an efficiency of about 1 % (for the $CO_2 \rightarrow HCOOH$ reduction reaction the Gibb's energy change is $\Delta G = 1.48 \, eV$). But when use is made of a p-GaAs electrode, the main reduction product is not formic acid but methanol [130], whereas CO is the main product on a p-CdTe cathode [131]. Thus, the process can be controlled by properly choosing the photocathode material.

The so-called "photo-Kolbe reaction" is similar to the Kolbe reaction well-known in organic electrochemistry. (The Kolbe reaction is the anodic oxidation of carboxylic acid anions, yielding saturated hydrocarbons.) For example, oxidation of acetate ions on an illuminated TiO_2 anode proceeds as per the scheme:

$$CH_3COO^- + h^+ \rightarrow \tfrac{1}{2} C_2H_6 + CO_2 \qquad (4.2)$$

and gives rise to the organic fuel – ethane.

On the whole, the studies into the photoelectrochemistry of organic substances on semiconductor electrodes have rather demonstrated the wide scope of this scientific trend, but have not yet lead to a well-founded final selection of particular practically important photoelectrochemical processes. This remains to be the task of the future.

4.3 Photoelectrodes made of Semiconductors with a Layered Crystal Lattice: Photointercalation and Photodeintercalation Processes

4.3.1 Introduction

The absence of photoanodes with an "optimum" set of photoelectric and electrocatalytic properties caused the scientists to find electrode materials among the classes of semiconductor compounds new for photoelectrochemistry. In the late 1970s Tributsch [132] suggested to consider the electrodes made of transition-metal dichalcogenides with a common formula TX_2, where T = Mo, W, Hf, Zr, Ru, Pt, and others, and X = S, Se, Te as objects of photoelectrochemical studies. The forbidden bandwidth of most compounds of this class comes to 1–2 eV, which is just suitable for solar energy conversion.

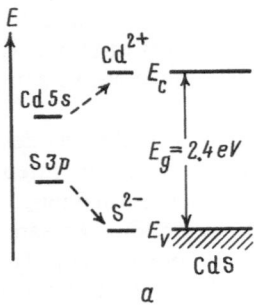

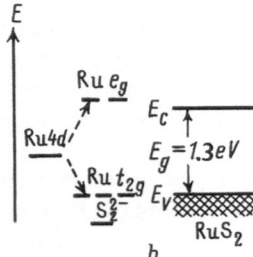

Fig. 63. Energy diagram of CdS (a) and RuS$_2$ (b) [132] Formation of energy bands of semiconductors from individual atomic levels is shown

Initially the strategical idea of this trend was formulated as follows. For a photoanode to be efficient and stable at the same time, it is necessary that the water photooxidation activation energy be as low as possible and the photocorrosion activation energy be as high as possible so that the former reaction predominates in the photocurrent. In the majority of considered compounds (in any case, in those compounds which are of prime interest for photoelectrochemical energy conversion) both the bands between which the electron transitions take place under the action of light – the conduction as well as valence bands – are composed of d-state orbitals of a transition metal (Fig. 63 b). In this they differ from sp-semiconductors like the chalcogenides CdS, CdSe, and others in which the interatomic bonds are largely ionic and the conduction and valence bands are in the main made by the orbitals, respectively, of the electropositive and electronegative components of the compound (Fig. 63 a). In the sp-semiconductors a hole weakens the interatomic bond by localizing on it and may lead to rupture. This underlies the anodic dissolution and corrosion processes. On the contrary, in d-semiconductors the formation of holes (say, as a result of d → d transitions upon illumination) affects only the quasi-nonbonding electrons and therefore does not cause the bonds to weaken. For this reason the d-semiconductors were assumed to be more stable to corrosion (and photocorrosion) than the sp-semiconductors.

At the same time, the holes formed upon illumination can participate in the reactions which do not affect the electrode material, say, in photoevolution of oxygen from water.[3] The more so, a favorable coordination of OH$^-$ (or H$_2$O) may be

---

[3] Compared to the mixed oxides considered in Sect. 3.5.4, the transition-metal dichalcogenides have a valence band of greater width (owing to the presence of a ligand like sulfur). Therefore, the value of anodic photocurrent would not be limited by an insufficiently large mobility of holes.

expected just on the surface of transition metals dichalcogenides, which ensures effective interfacial electron transfer without the intermediate formation of high-energy particles (for example, $OH^{\cdot}$ radicals); it is the intermediates that cause unjustified large energy losses at oxide photoanodes.

These hopes have yet been realized only partly. In fact, certain representatives of the considered class of semiconductors - RuS_2, $Ru_{1-x}Fe_xS_2$, PtS_2 - have proved to be stable against anodic photocorrosion. And photoevolution of oxygen from water (see Sect. 3.2) takes place on these semiconductors, even though the efficiency is small.[4] However, most of the transition-metal dichalcogenides as well as chalcogenides, say, of cadmium or zinc, decompose during anodic polarization in aqueous solutions upon illumination, with the only difference that here the anode decomposition product is not a free chalcogen (S, Se) on the electrode surface, but are soluble sulfates (or selenates). It is believed [133] that here the electrode material corrodes not just due to the photogeneration of holes and rupture of bonds in the crystal lattice, but under the chemical action of some anodic oxidation intermediate product in the solution. However, from the practical viewpoint this difference in the detailed corrosion mechanism is not of prime importance. Therefore, semiconductors - transition-metal dichalcogenides - are used as photoanodes not in the cells for photoelectrolysis of water, but in regenerative photoelectrochemical cells (see Chap. 6) where the electrodes are protected against corrosion by the redox system in the solution.

A very remarkable feature of many semiconductors of the considered class is that they have a layered crystal lattice.[5] The lattice is made up of closely packed atom layers positioned at relatively large intervals and bonded with each other by weak Van der Waals forces. This is the source of significant anisotropy of mechanical, electrical, and surface properties. The "layered" nature of the crystal lattice endows these compounds with the ability to intercalate, i.e., to implant alien atoms in interplanar spaces. Intercalation[6] may proceed under cathodic polarization with the aid of an external voltage source. Use of this is already made in producing electric secondary batteries with electrodes of "layered" compounds having metallic conductivity [134]. For example, intercalation of TiS_2 by lithium (with the formation of a compound $TiLi_yS_2$, where $0 \leq y \leq 1$) is a well reversible process and permits up to 480 Wh/kg energy to be stored. In the case of semiconductors, intercalation (as well as the reverse process - deintercalation, i.e., the process of removing the inserted species from the matrix material) may take place upon illumination under the action of the photopotential developed in the semiconductor itself. This opens a way of solar energy conversion with its simultaneous storage.

"Layered" semiconductors are generally synthesized from the gas phase; less often they are crystallized from the solutions in molten metals. The classification

---

[4] The authors of Ref. [132] see the cause of low efficiency in a too small value of the electrode photopotential (due to insufficient barrier height at the interface), which in its turn is caused by trapping of holes on the surface d-states pertaining to Ru atoms. It is not fully clear whether it is possible in principle to overcome this limitation.

[5] With these are grouped, for example, MoS_2, $MoSe_2$, WS_2, WSe_2, PtS_2, ZrS_2, $ZrSe_2$, HfS_2, and $HfSe_2$. Other dichalcogenides of transition metals - RuS_2 and FeS_2 - crystallize with the structure of pyrite.

[6] Sometimes the term insertion is also used in the literature.

of these materials and a review of their physicochemical and surface properties as compared to the pecularities of the crystalline and electronic structures can be found in Refs. [133, 135, 136].

4.3.2 Photoelectrochemical Behaviour of Semiconductors with a Layered Crystal Lattice

In the layered crystal lattice schematically shown in the inset (Fig. 64), the layer of transition-metal atoms (denoted by small black circles) is confined between two layers of chalcogen atoms (denoted by open circles). These closely packed "sandwiches" are put together in "stacks" where they are held by Van der Waals forces. The latter are relatively weak forces and therefore the crystal is readily cleaved in the direction parallel to the layers.[7]

The surface parallel to the layers (known as "Van der Waals surface" and denoted by the letter N in Fig. 64) features regular structure. Its chemical behavior depends on that the outer layer of chalcogen atoms (of sulfur, selenium, or tellurium) effectively screens the metal atoms responsible for electronic d → d excitation and for the formation of holes, from the interaction with the external medium (for example, the electrolyte solution). On this surface there are no unsaturated ("dangling") bonds; on the whole, it features reduced reactivity. In contact with an electrolyte, such a semiconductor electrode provides an example of "ideal" behavior.

The surfaces obtained by cleaving the crystal at an angle to the Van der Waals surface have entirely different properties. Here, the closely packed "sandwiches" terminate with "steps" (denoted by the letter R in Fig. 64); the unsaturated bonds

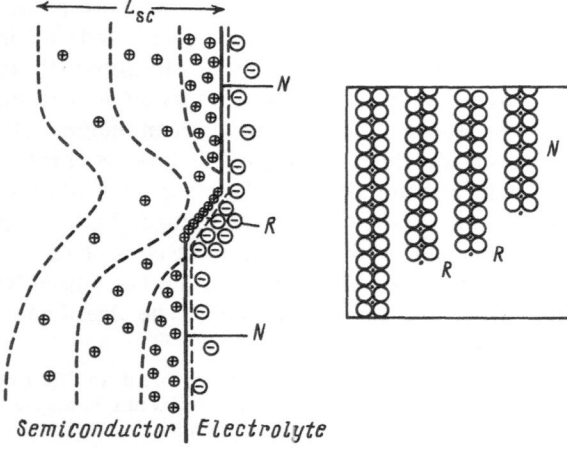

Fig. 64. Structure of double layer at the "layered" semiconductor/electrolyte interface [133]
Distribution of charge and equipotentials (dashed lines) in the depletion layer within the crystal are shown ,
The inset – schematic representation of "layered" crystal lattice

---

[7] The use of this fact is made in the experiment with "layered" semiconductors to renew their surface: an adhesive tape is first pasted on the surface and then removed, detaching thereby a thin layer of the material from the sample.

are not screened from the solution and act as active sites for chemical reactions. Thus, the surfaces rich in steps are particularly susceptible to photocorrosion [137]. The steps serve as recombination centers and as traps for electrons and holes. The local surface charge trapped at steps causes the space charge layer to become thinner at these places (Fig. 64). The increased surface recombination velocity impairs the photoelectrochemical characteristics of electrodes; therefore, chemical passivation techniques have been worked out to neutralize the ill effects of steps (see Sect. 6.2.3).

For obtaining maximum effectiveness in the processes that do not involve chemical changes in the electrode material (for example, in regenerative photoelectrochemical cells), it is recommended to make use only of "ideal" Van der Waals surfaces.

Earlier, in Sect. 1.4, it has been mentioned that, as on several other semiconductor electrodes, Fermi-level pinning on the electrode surface sometimes takes place on the "layered" semiconductors. This manifests itself in the shift of the Mott-Schottky plot upon illumination (Fig. 23) and also in that the photopotential is independent of the reversible potential of the redox system in the solution. The cause of this pinning may be the increased adsorption of ions (in particular, of I^-) on the electrode surface, particularly on the steps and defects, and/or high density of surface states. All this causes the potential drop in the Helmholtz layer to increase. Fermi-level pinning on the surface restricts the photopotential and, hence, the efficiency of the cells for light energy conversion.

4.3.3 Photointercalation and Photodeintercalation

In the process of intercalation, both the chemical composition of the electrode-matrix and its electron spectrum change. The intercalation scheme for a p-type semiconductor is shown in Fig. 65 a. The photoelectrochemical cell has a semiconductor photocathode and a metal (A) anode; an anode metal salt serves as electrolyte. When the photocathode is illuminated, A^+ ions discharge at the cathode, involving the photogenerated electrons, to form A atoms. In so far as A interacts with the semiconductor-matrix material it is energetically advantageous for A atoms to be inserted into the interplanar spacings (in the form of A_{int}) than to remain on the cathode surface.[8] The equation of reaction on a photocathode is:

$$TX_2 + A^+_{solv} + e^- \rightarrow TA_{int}X_2 + solv \tag{4.3}$$

(Here, *solv* stands for solvent.) As shown in the lower part of Fig. 65 a, desolvation takes place in the course of discharge and intercalation. (If the inserted atoms, however, retain part of their solvate shell so that the solvent molecules (denoted by S) enter into the semiconductor crystal lattice, then this causes increased corrosion of the semiconductor.) This is how a new redox system A^+_{solv}/A_{int} is formed. Its electrochemical potential level $F_{A^+_{solv}/A_{int}}$ is shown in Fig. 65. As the intercalation

---

[8] The same happens in electrochemical insertion of metals in metal electrodes, yielding intermetallic compounds and solid solutions [138].

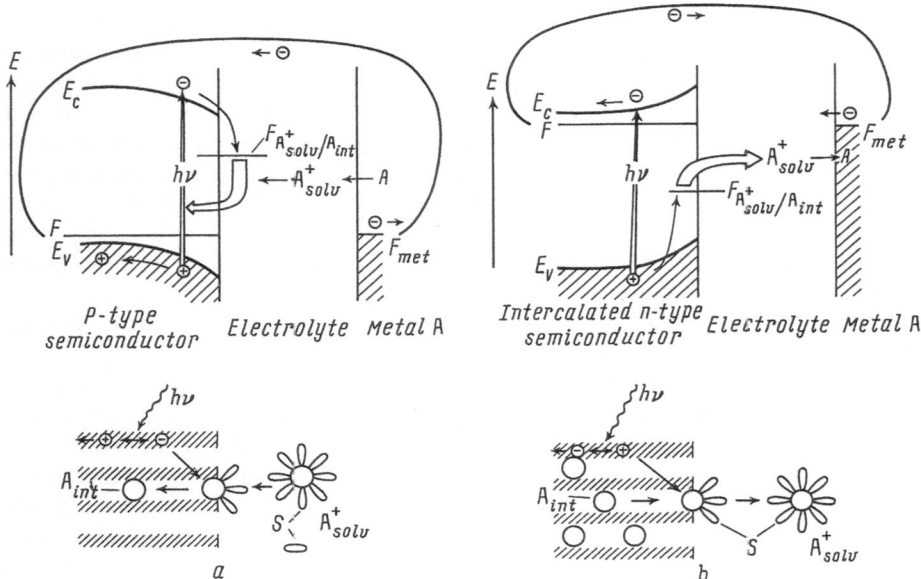

Fig. 65. Energy diagram (at the top) and schemes of processes of photointercalation of p-type semiconductor (a) and of photodeintercalation of n-type semiconductor (b) [136]
$F_{A^+_{solv}/A_{int}}$ – electrochemical potential level of the redox system (ions of metal in solution/atoms of implanted metal in the semiconductor); S – solvent molecule
Thick arrows denote processes of inserting A atoms into crystal lattice of cathode (a) and their escape into solution (b)

process proceeds, the location of this level and, together with this, the cathode potential change.[9] Here, two questions arise: first, how large can the magnitude of this shift be (since the value of electrode potential determines the thermodynamic possibility of intercalation)? And, second, to what extent can intercalation be carried out without losing the semiconductor properties of the matrix-electrode,[10] mainly the photoactivity which is the motive force of the electrode process? The answers to these questions should be found by experimentation. Unfortunately, the experimental data available to date are very limited.

Thus, the entire photoprocess in the cell amounts to dissolving the anode metal A and to forming on the cathode a new compound of variable composition:

$$TX_2 + A \xrightarrow{h\nu} TA_{int}X_2 \tag{4.6}$$

---

[9] Thus, in the enumerated system $TiLi_yS_2$, the potential reversibly shifts by 0.6 V when y changes between 0 and 1.
[10] Among the possible changes in the semiconductor properties of the matrix-electrode may be: change in the flat band potential and the band edges location on the surface; change in the forbidden bandwidth; variation in the bulk electrical conductivity and, hence, in the space charge layer thickness; and change in the band bending.

This causes accumulation of chemical energy whose value depends on the nature and amount of the inserted material A_{int}. On switching off light, the inverse process takes place:[11] the A_{int} atoms, by moving away from the semiconductor crystal lattice, ionize at the semiconductor surface, and the A^{+}_{solv} ions get reduced on the metal counter-electrode surface. When this happens, an electric current flows through the external circuit and voltage can be tapped off from the load resistor present in the circuit, as with discharge of a customary battery.

A variant of the above-considered process is shown in Fig. 65 b: upon illumination, the pre-intercalated n-type semiconductor photoanode gives up the accumulated species, A_{int}.

Finally, a more trivial process of "passive photointercalation" of the metal cathode of a photoelectrochemical cell is possible under the action of photopotential developed by the semiconductor photoanode. The material of the latter does not directly participate in intercalation.

The thermodynamic consideration of the intercalation process [135] leads to the following approximate expression for the "dark" potential of the intercalated electrode:

$$\varphi = \varphi_0 + By - \frac{mkT}{e} \ln \frac{y}{y-1} \tag{4.5}$$

Here, $y = [A_{int}]/[A_{int}]_{max}$ is the degree of filling up "vacant places" in the matrix by the inserted substance, B is the factor that describes the nature of interaction between the particles of this substance (e.g., $B < 0$ when there are mutually repelling particles), m is the number of sorts of "places" for intercalation ($m = 1$ or 2). In Fig. 66 the A–D curve presents the dependence of potential on the degree of filling, y. Here, φ_A is the initial (prior to illumination) potential of the electrode which is in equilibrium with the counter-electrode in the cell. This potential is conditionally taken equal to 0.

Upon illumination, the electrode potential becomes more positive by a value φ_{ph} which is determined by Eq. (2.5). In Fig. 66 (where, by way of example, φ_{ph} is taken equal to 0.5 V) this is represented by segment A–B. The intercalation process starts and in the course of this process (see Eq. (4.5)) the potential of the illuminated electrode decreases (segment B–C) until the potentials of both cell electrodes become equal. The light is then off; the potential of the intercalated electrode shifts from C to D. The chemical energy accumulated in the electrode is proportional to the hatched area A–C–D. When the cell electrodes are connected to an external load, the deintercalation process proceeds spontaneously (i.e., the electrode discharges). To this corresponds segment D–A. In practice, because of the not quite "ideal" behavior of a photoelectrode (the reason being, in particular, the effect of the inserted substance on the matrix-semiconductor properties), the photopotential is found to be less than its limiting value (to which corresponds the

---

[11] Therefore, the external circuit must include a rectifying element – diode (not shown in Fig. 65) in order to prevent spontaneous discharge of the photoelectrode during photointercalation in case of accidental dim-out.

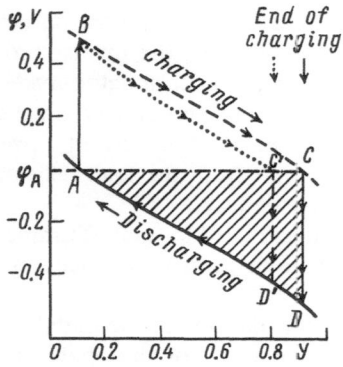

Fig. 66. Schematic representation of variation of p-type semiconductor electrode potential when light is switched on (A-B), during charging (intercalation) with light on (B-C), when light is off (C-D) and during discharging (deintercalation) in the dark (D-A) according to Eq. (4.5)
Calculations have been performed for B = -0.35, m = 2, $y^{max} = 0.92$. Dotted line shows the case of "nonideal" behavior [135]

case $\gamma > 1$ in Eq. (2.5)). The degree of intercalation, y, and hence the energy stored in the electrode is also less than the limiting values. To this case corresponds the cycle A-B-C'-D'-A shown by the dotted line in Fig. 66.

On the strength of Fig. 66 one can formulate some of the limitations on the selection of the cell anode material (and, hence, of the substance inserted in the photocathode). First, the potential difference between the discharged (i.e., not intercalated) semiconductor cathode and the anode must not exceed 0.5-1.0 V because, for a forbidden bandwidth of 1-2 eV, it is not possible to obtain larger values of photopotential. (That is why it is not possible to use alkali metals for intercalation, as this is done, e.g., in TiLi$_y$S$_2$-based secondary cells, though they enable very high values of specific stored energy to be obtained; their reversible potentials lie in the range of -2.5 to -3.0 V, which cannot be attained as semiconductor electrode photopotential, and thus rules out spontaneous intercalation.) Second, intercalation proper should be an "uphill" stage of the entire reaction, i.e., should occur spontaneously only under the action of light energy but not in the dark. Copper, silver, and indium which satisfy both these conditions, are generally used as anodes.

In a like manner is described the photodeintercalation process of a pre-intercalated n-type semiconductor which acts as photoanode in the cell. Under the action of photogenerated holes, the substance inserted in the semiconductor crops out and ionizes (Fig. 65 b). In order to use Fig. 66 for describing this process, "revolve" the figure around the horizontal line A-C so that the segment B-C is below the segment A-D. The counter-electrode potential should be taken more negative than the potential of the intercalated electrode so that intercalation (discharge) will occur spontaneously in the dark and photodeintercalation (charging) will take place under illumination.

In principle, use can be simultaneously made in the same cell of photointercalation of a p-type semiconductor and of photodeintercalation of a n-type semiconductor. This permits to obtain larger values both of photovoltage across the cell and stored energy.

The efficiency of a photoelectrochemical cell in which use is made of the intercalation phenomenon is expressed as:

$$\eta = \eta_g(y)\,\eta_{int}(y) \tag{4.6}$$

Here, η_g is the efficiency proper of the photocurrent generation stage (it is determined by analogy with the efficiency of a regenerative photoelectrochemical cell, see Sect. 6.1), and η_{int} characterizes the efficiency of the photointercalation stage and of the stage of subsequent utilization of the stored energy. In the general case, both these quantities are dependent on the degree of intercalation, y.

Some of the semiconductors that have been studied by now and can be used as matrix-electrode, and the intercalable substances suitable for these semiconductors are listed in Table 4.1. Besides the dichalcogenides of transition metals, the table includes compounds of other classes. These are first of all compounds with the formula M_6PX_5Hal (where M stands for metal, X for chalcogen, and Hal for halogen). During deintercalation (by converting into a compound of the type $M_{6-x}PX_5Hal$) they retain their semiconductor properties including photosensitivity, and are capable of undergoing photointercalation. (For example, photointercalation of $Cu_{6-x}PS_5I$ in the acetonitrile solution of Cu^+ proceeds until a stoichiometric compound Cu_6PS_5I is obtained.)

In summarizing, it must be mentioned that the studies into the entire range of diverse problems associated with the photoelectrochemistry of "layered" d-semiconductors are far from completion. In particular, it is not yet fully known to what extent an electrode can be intercalated without appreciably impairing its semiconductor properties, first of all photosensitivity; corrosion and other phenomena are also not well studied.

Table 4.1. Photoelectrochemical intercalation and deintercalation systems [136]

Matrix-semi-conductor	E_g, eV	Type of conductivity	Solvent	Intercalable substances	Type of process	Limitations
$ZrSe_2$	1.05–1.22	p	Acetonitrile	Li	Photointercalation	Excessive intercalation results in quasi-metallic conductivity
$HfSe_2$	1.13	n	Water, acetonitrile	Na	Photodeintercalation	
ZrS_2	1.68	n	Acetonitrile	Na	Photodeintercalation	
HfS_2	1.96	n	Water	Na	Photodeintercalation	Low conductivity
$FePS_3$	1.5	p	Acetonitrile	Li	Photointercalation	Small quantum yield
InSe	2	n	Propylene carbonate	Cu	Photodeintercalation	
Cu_3PS_4	2	p	Acetonitrile	Cu	Photointercalation	
$Cu_{6-x}PS_5I$	2.05	p	Acetonitrile	Cu	Photointercalation	
TiO_2 (B) [a]	3	n	Water	H	Photodeintercalation	High dark current

[a] Specially prepared TiO_2 with a "layered" lattice.

Chapter 5
Solar Energy Conversion into Chemical Energy.
Suspensions and Colloidal Solutions

In recent years an extensive exchange of ideas has taken place between photoelectrochemistry, photocatalysis, and photobiology. As a consequence, great interest has aroused in the photochemical behavior of microheterogeneous systems - suspensions, colloidal solutions, mycelles, and vesicles - which enable directed photochemical processes to be carried out with a greater efficiency than it has been possible in homogeneous systems. The use of microheterogeneous semiconductor systems holds much promise. Though the important advantage of macroscopic electrochemical cells, namely the possibility of spatial separation of electrolysis products is offset, the microheterogeneous systems combine photosensitivity inherent in semiconductors with large photocatalytic activity typical of the systems with a developed surface.

Increasing the true surface of a semiconductor is equivalent to decreasing the true photocurrent density: this means, first, a decrease in electric losses due to overvoltage of the electrochemical reactions, and, second, a decrease in the relative rate of photocorrosion which (see below, Fig. 78) proceeds more intensively at large photocurrent densities. Therefore (leaving aside the question of "mechanical" instability of microheterogeneous systems caused by the coagulation of colloidal particles or by sedimentation of suspensions), the microheterogeneous semiconductor systems are expected to be chemically more stable compared to macroscopic cells with photoelectrodes made of the same materials.

Finally, an advantage of the systems with very small particles (whose size is less than the light wavelength) is that they do not scatter light; lack of reflection and scattering losses increases the quantum yield of the photoprocess.

Ample literature is dedicated to microheterogeneous semiconductor systems (see, for example, reviews of Ref. [46, 139] and the book cited under Ref. [2] (Chaps. 3, and 7 through 10)). Unlike other fields considered in our book, this field is being developed very intensively. Here, many things are still unsettled and often no unequivocal conclusion can be made about the mechanism of processes in particular microheterogeneous systems. Therefore, below we shall mention only the fundamentals of the photoelectrochemical action of these processes and the cited examples will be rather illustrative.

5.1 Mechanisms of Photoelectrochemical Processes in Microheterogeneous Systems

Microheterogeneous systems can be compared, on the one hand, with the homogeneous photochemical systems enumerated in the introduction, and, on the other hand, with the macroscopic photoelectrochemical cells (to which this book is primarily dedicated).

For the water photodecomposition reaction, Fig.67 gives a schematic representation of the type and basic stages of photoprocesses that occur in homogeneous and microheterogeneous systems. To start with, we shall compare these with homogeneous systems. Figure 67a shows the energy diagram typical for classical photochemical studies of a homogeneous system containing a dye-sensitizer (S and S* denote respectively its ground and excited states) and an acceptor-type supersensitizer (R) in an aqueous solution. The light-excited form of the sensitizer donates an electron to the supersensitizer; the products of this reaction, S^+ and R^- respectively oxidize and reduce water forming O_2 and H_2 and regenerating the starting reagents S and R:

$$2R^- + 2H_2O \quad \rightarrow \quad H_2 \ + 2OH^- + 2R \tag{5.1a}$$

$$2S^+ + H_2O \quad \rightarrow \quad \tfrac{1}{2}O_2 + 2H^+ \ + 2S \tag{5.1b}$$

Among sensitizers, extensive use is made, of the bipyridyl complex of Ru(II), $(Ru(bpy)_3^{2+})$; methylviologen can act as a supersensitizer.

The impossibility of discarding completely the deactivation reaction $(S^+ + R^- \rightarrow S + R + heat)$ caused scientists to look for, as already mentioned in the introduction, heterogeneous systems capable of making the charge separation process irreversible. An example of this is a semiconductor/sensitizer solution system (Fig.67b) where a semiconductor particle acts as a supersensitizer. The forbidden

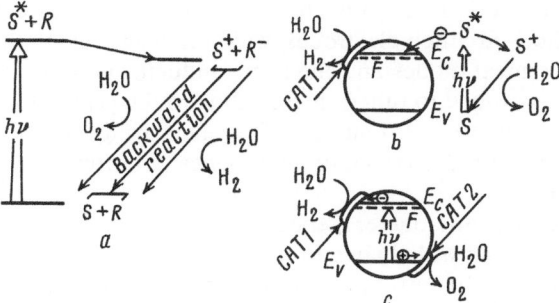

Fig. 67. Types of photoprocesses in illuminated photoactive systems for water decomposition [139]
(a) homogeneous system with sensitizer S and acceptor R; (b) microheterogeneous system with sensitizer S and particles of semiconductor that does not absorb light; (c) microheterogeneous system with particles of semiconductor which absorbs light
E_C and E_V are the energies of conduction band bottom and valence band top in semiconductor particle; CAT 1 and CAT 2 are respectively the islets of H_2- and O_2-evolution catalysts on the particle surface; S and S* stand for the ground and excited state of sensitizer

band in the semiconductor makes the electron transfer from the excited sensitizer to the semiconductor conduction band irreversible (cf. photosensitization of macroscopic semiconductor electrodes, Sect. 3.5.3). Here it must be stressed that in the case shown in Fig. 67b, light is absorbed not in the semiconductor but in the solution. As before, water gets oxidized in the course of its homogeneous reaction with S^+, but water reduction now takes place on the semiconductor particle surface (more precisely, on the islets of a catalytically active metal, see below). Such a scheme for photodecomposition of water with the use of $Ru(bpy)_3^{2+}$ has been described by Grätzel [139].

Finally, Fig. 67c illustrates the case of particular interest to us, i.e., the case when the semiconductor particle itself acts as photosensitizer for the process owing to the occurrence of photogenerated primary excited states – non-equilibrium electrons and holes which further participate in electrode reactions of the type in Eq. (2.7). Necessary selectivity towards these reactions can be created, as suggested in Ref. [139], by putting two types of catalyst-islets on the surface of each particle; for example, in the photodecomposition of water, the water reduction reaction with conduction band electrons can be catalyzed by platinum group metals, and the water oxidation reaction with holes, by ruthenium dioxide (an oxide with metallic conductivity).

As a logical link between the macroscopic photoelectrochemical cells and the microheterogeneous systems may serve the so-called "photochemical diode" suggested by Nozik [140]. It represents two – semiconductor and metallic – tablets placed together and connected with each other by an ohmic contact. The linear dimensions of this model may vary from some millimeters to a micron. A set of such "diodes" is placed in the agitated electrolyte solution and illuminated. The operation of a photochemical diode is considered as operation of a short-circuited galvanic pair: the same reactions occur on its "electrodes" as on the electrodes of a common electrochemical cell. For example, when $SrTiO_3$ and Pt are combined, photodecomposition of water takes place in aqueous solution and oxygen and hydrogen are evolved, respectively, on the surface of strontium titanate and platinum. In fact, a photochemical diode is none other than an electrochemical cell (see for example Fig. 22) with short-circuited electrodes, which is "split" along the electrolyte. Indeed, the photochemical diodes and the semiconductor particles in microheterogeneous systems are assumed to function according to the principle of photosynthetic cells (see the scheme of Fig. 25) but not of regenerative cells, since in the latter useful work cannot be obtained owing to the absence of current leads, and light energy changes into heat.

So, semiconductor particles in a microheterogeneous system may be represented as microgalvanic pairs. But what will change in the operation of the above-considered photochemical diode if its dimensions are progressively decreased? To answer this question, it is necessary to compare the semiconductor particle radius r with other characteristic dimensions: space charge layer thickness (which by the order of magnitude is equal to the Debye length L_D), light penetration-depth in the semiconductor (α^{-1}, where α is the absorption coefficient, see Sect. 1.2), and the diffusion length of minority carriers, L_p. The photochemical behavior of a semiconductor particle depends on the relation between the enumerated quantities.

Let us now consider in more detail the stages of the process shown in Fig. 67 c, namely the separation of photogenerated charges in a semiconductor particle and their transfer through the interphase boundaries (by considering the example of a n-type semiconductor) [141, 142].

First we shall refer to "large"-size particles such that:

$$r \gg L_D, \quad r \gg \alpha^{-1} \tag{5.2}$$

Recall that the space charge layer thickness is given by Eq. (1.16); for a n-type semiconductor at room temperature it is expressed as:

$$L_D = 84.6(\varepsilon_{sc}/n_0)^{1/2} \quad (cm) \tag{5.3}$$

where ε_{sc} is the relative dielectric permittivity of the semiconductor and n_0 is the bulk concentration of electrons (cm^{-3}).

In the given context the concepts of a "large" and "small" particle are relative. For example, at $\varepsilon_{sc}/n_0 = 10^{-14}$ (a relatively weakly doped semiconductor) $L_D \simeq 10^{-5}$ cm such that a particle of radius 1 μm is considered "large" and of radius 10 nm, "small". At a much higher doping level ($\varepsilon_{sc}/n_0 = 10^{-16}$ to 10^{-17}) a particle of radius $r = 10$ nm can no more be regarded "small". The absorption coefficient α usually amounts to $10^3 - 10^5$ cm^{-1}; therefore, for "strongly absorbable" light at $r = 1$ μm, the second condition, Eq. (5.2), is also fulfilled, whereas for "weakly absorbable" light this condition is not satisfied.

In a "large" particle, separation of photogenerated carriers takes place by the same mechanism as in a macroscopic electrode (cf. Sect. 2.1) whose basis is an electric field in the space charge layer (for definiteness, we shall consider the case of depletion layer). Figure 68A shows the energy diagram of a particle conditionally represented as a membrane of thickness 2r, surrounded by solution on both sides. The minority carriers – holes – in the depletion layer field move towards the particle surface and can pass into the solution containing a suitable reducing agent Red$_1$. The majority carriers – electrons – go into the particle bulk and accumulate there, because in the case of a separately taken semiconductor particle there is no effective "drain" for the majority carriers (in a short-circuited cell the ohmic contact together with the external circuit and a counter-electrode serve as "drain"). Since a particle as a whole remains electroneutral, the accumulation of electrons in the bulk tends to cease the transfer of holes into the solution and a steady state is established. In this state the carriers' photogeneration rate is equal to their recombination rate. Illumination causes unbending of bands and the field in the depletion layer decreases, which tends to reduce the charge separation efficiency. In effect, the considered situation for a separate particle is equivalent to the case of an illuminated semiconductor electrode under open-circuit conditions.

So far as the depletion layer is retained in a particle, the transfer of majority carriers into solution is hampered. It is true that even in the presence of a depletion layer the electrons can go into solution by diffusion occurring against the electric field, if the latter is not very strong (see the remark on p. 32). In photoelectrochemical cells, simultaneous transfer of electrons and holes from the elec-

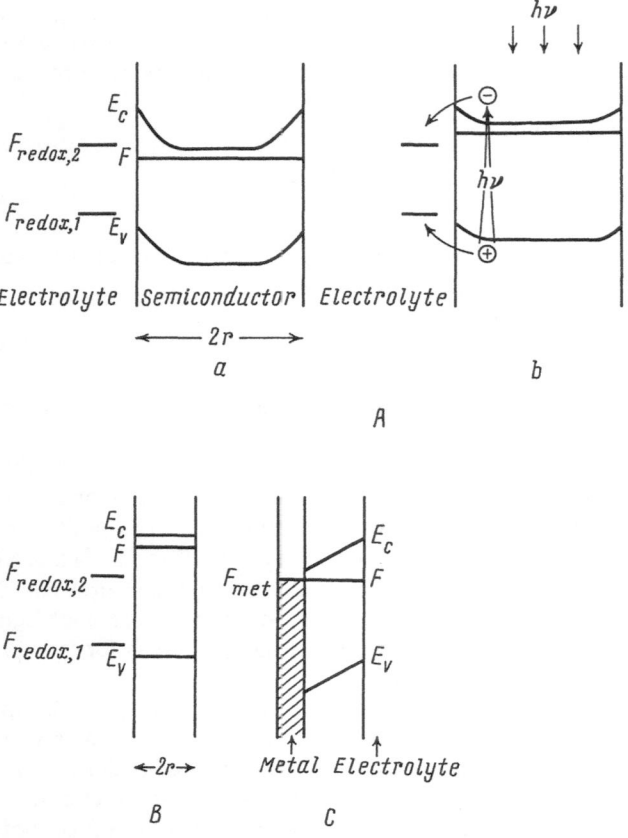

Fig. 68. Energy diagram of particle of n-type semiconductor in electrolyte solution in the dark (a) and when illuminated (b) at $r \ll \alpha^{-1}$
In the solution are shown the electrochemical potential levels of two redox systems. A: "large" particle, $r \gg L_D$; B, C: "small" particle, $r \ll L_D$; C – particle with a coated metal-catalyst
The band edges (E_C, E_V) are assumed to be pinned on the semiconductor surface

trode into solution causes the quantum yield to decrease. But in the case of suspension particles, such a transfer may lead to the occurrence of the desired photosynthesis reaction if the electron flow and the equal hole flow are directed onto different reagents in the solution: electrons – for the reduction of Ox_2, and holes – for the oxidation of Red_1. (If, on the contrary, both flows are directed onto the same system Ox-Red in the solution, then the illuminated particle virtually functions under the conditions of a regenerative cell but with no current pick-off, which, as mentioned earlier, leads only to efficient "electrochemical recombination" of photogenerated carriers and to conversion of their energy into heat.) Therefore, one must aim at directing the minority and majority carrier flows in such a manner that they intersect the semiconductor/solution interface at different sites. Otherwise, the most probable result will be the recombination of electrons and holes at the surface.

In principle, it is possible to spatially separate electrons and holes on the particle surface (and, hence, to direct them onto different reagents in the solution), creating different-in-reactivity sites on the surface. The simplest form of such a surface spatial inhomogeneity appears when the surface is unevenly illuminated. In a "large" particle when $r \gg \alpha^{-1}$ such different portions are the illuminated (frontal) and unilluminated (rear) sides of the particle. As is known, in an irregularly illuminated semiconductor electrode, a potential difference appears between the illuminated and unilluminated portions of the electrode under open-circuit conditions (this has been considered in detail in Sect. 10.1 of Ref. [1]). As a result, microgalvanic pairs are formed. In a n-type semiconductor the illuminated portions are local anodes on which the oxidation reactions proceed with the participation of holes, and the unilluminated portions are local cathodes where the reduction reactions occur with the participation of conduction band electrons. (In p-type semiconductors, illuminated portions act as cathodes and unilluminated portions are anodes.) Note that in a "small" (in the sense of $r \ll \alpha^{-1}$) particle, light is uniformly absorbed in the entire bulk and the difference between the "illuminated" and "unilluminated" sides disappear (just that very case is shown in Fig. 68).

A more effective way is to control the reactivity by putting islets of a catalytically active metal on the semiconductor surface, which selectively accelerates exactly the desired reaction. Let us consider the energy diagram of a semiconductor particle with a catalytically active metal on its surface. Suppose that the difference in the work functions of the semiconductor and the metal is such that an accumulation layer appears at the metal/semiconductor interface (Fig. 69). This contact is ohmic and permits the majority carriers – electrons – to readily leave the semiconductor particle and to pass into the metal and, therefrom, into the solution. (For the inverse relation of work functions, i.e., when a depletion layer is formed at the semiconductor/metal contact, the majority carriers can hardly escape from the semiconductor into the metal and the minority carriers escape readily, on the contrary.)

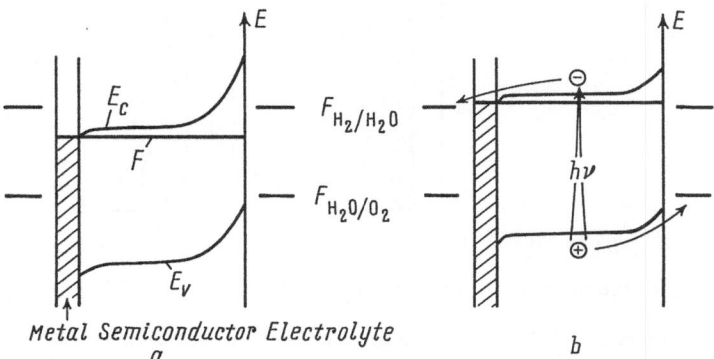

Fig. 69. Energy diagram of semiconductor particle with a metal-catalyst on the surface (case of ohmic contact)
(a) in the dark; (b) when illuminated

The examples of metal-catalysts that form ohmic contact with a semiconductor are platinum metals (Pt, Rh, and Ru) on TiO_2 and $SrTiO_3$, which catalyze the hydrogen evolution process. Strictly speaking, only ruthenium as such has a work function close to those of TiO_2 and $SrTiO_3$; therefore, it should form ohmic contact with these semiconductors. Rhodium and platinum, in particular, have a much higher work function; therefore, a depletion layer with a Schottky barrier height of up to 1 eV is formed in the enumerated semiconductors at the contact with these metals. But as already indicated in Sect. 3.2, dissolution of hydrogen in platinum and rhodium decreases the work function of both the metals up to approximately the work function of ruthenium (and also of TiO_2 and $SrTiO_3$) as viewed in Fig. 70A. Therefore, under the conditions of cathodic photoevolution of hydrogen, i.e., in the hydrogen atmosphere the contacts of platinum metals with TiO_2 and $SrTiO_3$ become ohmic as conditionally shown in Fig. 70B. Note that in such a system the hydrogen photoevolution process should be autocatalytic: its occurrence boosts the catalyst action, which in its turn accelerates the process, and so on.

The RuO_2/CdS system as well as the Pt/p-InP system discussed in Sect. 3.2 are examples of different kind: a Schottky barrier is formed at the semiconductor/catalyst contact, which promotes the interfacial transfer of minority carriers.

But it is only a half job to select electric properties of the catalyst/semiconductor contact, which will ensure easy transfer of the desired type of carriers from the semiconductor particle onto the catalyst. No less important is to select the chemical nature of the catalyst so that, first, the catalyst will enhance only the desired reaction and, second, only the direct but not the inverse reaction will be accelerated. For example, RuO_2 catalyzes the anodic evolution of O_2 and cathodic evolution of H_2 from water, but has little effect on the cathodic reduction of O_2. On the

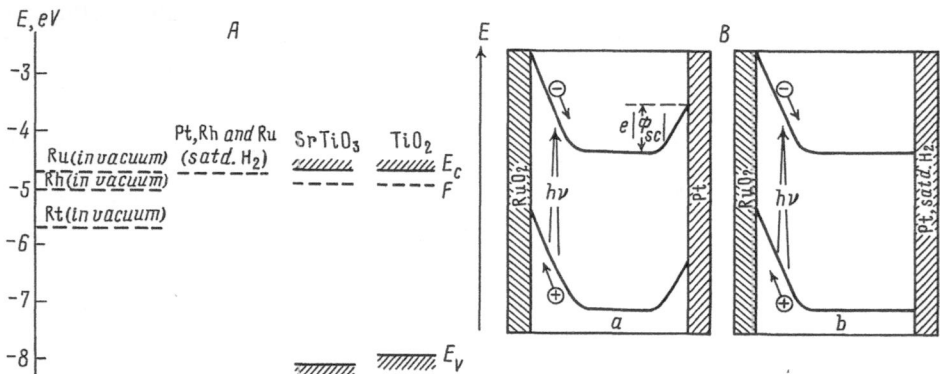

Fig. 70. Effect of hydrogen absorption on the properties of micro-deposits of platinum metals on semiconductors
A - comparison of work functions of Pt, Rh, and Ru in vacuum and in hydrogen atmosphere and energy diagrams of $SrTiO_3$ and TiO_2; B - schematic representation of TiO_2 semiconductor membrane with Pt and RuO_2 catalyst films deposited on opposite sides; a - in vacuum, b - in the atmosphere of hydrogen. The change in the nature of TiO_2/Pt contact due to hydrogen absorption is shown (reproduced from Ref. [119])

contrary, platinum accelerates both the evolution and ionization of oxygen. Therefore, when use is made of a platinum catalyst in microheterogeneous systems for photosplitting of water, one is particularly keenly faced with the problem of fast removal of products from the reaction zone. In fact, as soon as the solution gets saturated by the liberating gases, reduction of O_2 replaces the evolution of H_2 as a cathode partial reaction and the water-splitting efficiency decreases. (About the interaction of two catalysts simultaneously present on the same particle, see Sect. 5.3.)

Let us now turn our attention to small-size (in the sense that $r \ll L_D$), i.e., colloidal, particles. In these particles, there is no space charge and the band edges have the same energy at all points of the particle (Fig. 68B). It is significant that this energy does not change even upon illumination (in the considered simplest case of band edge pinning at the surface, and also in the absence of some subtle effects which will be discussed in Sect. 5.2). Insofar as there is no electric field within the particle[1], there is no immediate ground for separating the light generated electrons and holes. The mechanism of carrier transportation in a particle also changes. Now, the carriers can be transferred only by diffusion under the action of the concentration gradient. The latter appears if any portion of the surface acts as a "drain" for electrons or holes. Therefore, the local concentration of carriers in the neighborhood of this portion decreases. By the order of magnitude the carrier diffusion transit time over the particle equals (see Chap. 3 in Ref. [2]):

$$t \simeq r^2/\pi^2 D \qquad\qquad\qquad (5.4)$$

Here, D is the diffusion coefficient of carriers. At $r = 10$ nm and $D = 10^{-2}$ cm^2/s, t equals 10^{-11} s; at $r = 1$ μm, $t = 10^{-7}$ s. These values of t are to be compared with the characteristic time of bulk recombination in semiconductors ($\simeq 10^{-7}$ s). Whence it follows that in "small" particles the recombination of carriers generated in the bulk may be neglected; in "large" particles, this cannot be done. That is why the presence of an electric field is not a necessary condition for the charges in "small" particles to separate. At the same time, in "small" semiconductor particles it is significantly important to ensure, by using catalysts, fast and selective transfer of electrons and holes from the semiconductor into the solution.

5.2 Microheterogeneous Systems: Fundamentals of their Preparation and Study Methods. Special Effects

Microparticles of semiconductors are prepared mainly by mechanical grinding or chemical synthesis. In the former case, dopants are introduced into the semiconductor beforehand, i.e., in obtaining the starting material, and in the latter case, in

---

[1] Given spatial chemical inhomogeneity, for example if part of the semiconductor surface is occupied by a metal-catalyst islet, there exists a constant electric field within the particle and the band edges are a linear function of the coordinate, as shown in Fig. 68C.

the course of synthesis proper. And the below given information on the methods of preparing suspensions and colloids is purely illustrative.

Thus, small particles of TiO_2 (with a crystalline structure of anatase) are prepared by hydrolyzing Ti(IV) compounds, e.g., $TiCl_4$ or titanium tetraisopropyloxide in aqueous solutions, and also by hydrolyzing vapors and aerosols, by high-temperature oxidation of $TiCl_4$, and by other methods (see Chap. 8 in Ref. [2]). The obtained particles are generally tens and hundreds of nanometers in size; they can be made conductive, if necessary, by introducing a few tenths of a percent of Nb in the bulk of TiO_2. (Recall that pure titanium dioxide is an insulator.) For this purpose, Nb_2O_5 was coprecipitated with TiO_2.

Catalytically active spots (in particular, the so-called "bifunctional catalyst" represented in Fig. 67 c) were obtained on TiO_2 particles by adding small amounts of RuO_2 and Pt. The former is introduced in the course of co-coagulation of RuO_2 and TiO_2 colloids, and the latter, upon adsorption of ultradisperse sol of Pt on TiO_2 particles or photoelectrochemical platinization (by illuminating TiO_2 suspension in the H_2PtCl_6 solution).

The CdS particles are synthesized in a solution, for example, from $Cd(ClO_4)_2$ and NaHS.

In certain cases the colloidal particles are stabilized against coagulation by using surface-active substances like polyvinyl alcohol. But here it must be kept in mind that organic stabilizers may get oxidized by photogenerated holes, which complicates the interpretation of the results of photoelectrochemical studies. Sometimes, a semiconductor material is deposited on the inert-carrier particles (SiO_2 [143], montmorillonite [144], and others). Semiconductor suspensions can be sensitized to a more long-wave light than the light absorbed by the semiconductor proper using adsorbed dyes, i.e., in the same manner as "macroscopic" electrodes (see Sect. 3.5.3).

Table 5.1. Methods in the characterization of semiconductor dispersions and in the study of their photoreactions (Chap. 7 in Ref. [2])

Property/reactions	Technique
Characterization	
Size and polydispersity	Low-angle light scattering; X-ray; electron microscopy
Concentration	Atomic absorption spectroscopy
Surface area, porosity	Gas adsorption such as BET
Charge	Micro- and photoelectrophoresis; slurry electrodes
Bandgap	Photoacoustic, diffuse reflectance spectroscopy
Surface acidity or basicity	IR spectroscopy, thermogravimetry; temperature-programmed desorption
Metal and metal oxide deposits	Electron spectroscopy (ESCA-, Auger-); transmission electron microscopy
Reaction	
Adsorption or desorption of gases	Photoconductivity; electron spin resonance; dynamic and mass spectrometry; Hall effect
Intermediates detection and mechanism	Flash-photolysis; ESR-spin traps; product analysis in static and flow reactors; photoelectrochemical cells with macrosemiconductor electrodes

A set of chemical, physical, optical, and structural methods is employed to determine different characteristics of microheterogeneous systems. Some of these methods are listed in Table 5.1.

Without dwelling on the essence of most of these methods (briefly it is described in Chap. 7 of Ref. [2]) we shall mention only the suspension electrode method. This method involves current take-off from the suspension particles through the use of a current-collector metal electrode, as shown in Fig. 71. For example, in the illuminated aqueous TiO_2 suspension, an anode photocurrent of the order of tens of a microampere can be obtained by using a platinum mesh collector electrode [145]. With such an electrode a potential (different from the stationary one) can be imposed on the suspension particles and the "current-potential curves" of the type obtained in photoelectrochemical cells with "macroscopic" electrodes can be drawn. An analysis of these curves (by making allowance for specific pecularities of the suspension electrode) provides a means for estimating the flat band potential, photocurrent onset potential, and other characteristics of particles. The suspension electrode theory [146] makes it possible to compute the current-potential curve. It has been found that this curve, in case of rapid mass transfer, is described, as for a common electrode reaction, by the Tafel law, even though the mechanism of the process is complex. (It must be remembered that upon collision with the collector electrode, different particles give or take unequal number of electrons.) Also, if the process rate is limited by the mass transfer, then quite simple relationships are sometimes observed. Thus, if the current-collector is a rotating disc electrode, then the limiting current is proportional to the square root of its rotational speed, as in the case of electrochemical reactions occurring with the participation of ions or molecules (see, for example, Ref. [147]).

The flat band potential of particles (and, hence, the band edge location on the surface, $E_{C,s}$ and $E_{V,s}$) can be determined without using a special electrode, but with the aid of a certain "reference" redox system in the solution, which is well reversible and has a known standard potential. An example of this system is methylviologen MV^{2+} (often used as a mediator of homogeneous photochemical reactions). Its standard potential $\varphi^0_{MV^{2+}/MV^+}$ equals $-0.45\,V$ and is independent of the solution pH. Therefore, when the pH of the methylviologen solution containing an oxide semiconductor suspension is varied, the semiconductor flat band potential φ_{fb} gets shifted (see Eq. (1.23)) relative to $\varphi^0_{MV^{2+}/MV^+}$. When the semiconductor Fermi level F exceeds the methylviologen electrochemical potential level F_{MV^{2+}/MV^+}, the reduction reaction of methylviologen starts with the involvement of electrons photogenerated in the semiconductor (the holes are generally consumed in the photo-

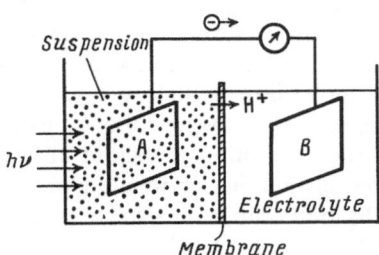

Fig. 71. Photoelectrochemical cell with a suspension electrode
A – collector-electrode; B – counter-electrode; the left (illuminated) compartment of the cell contains suspension of semiconductor particles

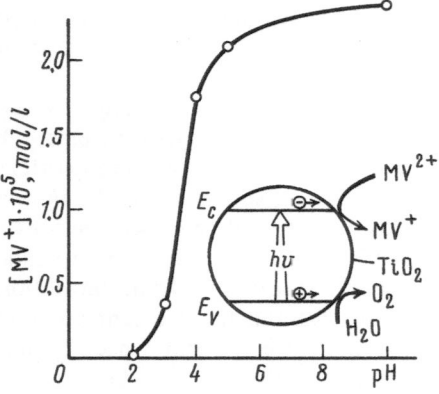

Fig. 72. Effect of solution pH on the yield of methylviologen cation-radical MV^+ when the colloidal solution of TiO_2 is illuminated (Chap. 3 of Ref. [2])
Concentration: $TiO_2 - 0.5$ g/l, methylviologen $(MV^{2+}) - 10^{-3}$ mol/l
The inset shows the scheme of the process

oxidation of water). The appearance of a characteristic band (with a maximum at 602 nm) belonging to MV^+ in the absorption spectrum of the solution marks the onset of this reaction (and, hence, the fulfilment of the condition $F > F_{MV^{2+}/MV^+}$). The pH dependence of the rate at which the reduced methylviologen accumulates in the aqueous TiO_2 suspension is shown in Fig. 72. From this figure it was found that $\varphi_{fb} = -0.12 - 0.059\,pH$, V (against a normal hydrogen electrode). This is how the location of F is determined. Thereafter, E_C, E_V, $E_{C,s}$, and $E_{V,s}$ are found in a usual manner (see Sect. 1.4, p. 24) and a complete energy diagram for the particle in solution is constructed.

We shall now highlight certain distinguishing features of the behavior of semiconductor dispersions, essentially caused by the small size of their particles. First of all, particles of tens or hundreds of nanometers in size behave practically as insulators. In fact, even if in a semiconductor the concentration of fully ionized donors is not so small, e.g., 10^{15} cm^{-3}, nonetheless, its one particle of radius 50 nm will on the average contain less than one conduction electron, i.e., it will be an insulator. Therefore, a situation is quite probable – when upon absorption of light the number of photogenerated majority carriers far exceeds the number of the carriers in the dark. And the accumulation of non-equilibrium carriers causes an unusually strong shift in the quasi-Fermi level for the majority carriers in the particle when illuminated.

Another manifestation of quantum size effects in small semiconductor particles is the increase of the forbidden bandwidth E_g and the associated with this "blue" shift of the intrinsic light absorption threshold [148]. This effect was observed in the colloidal particles (with radius of the order of several nanometers) of ZnS, CdS [149], HgSe, PbSe, CdSe [150], and other materials. The change in the absorption threshold as compared to "macroscopic" semiconductors may be as large as several tenths of an electron volt; it decreases with growing colloidal particles and finally disappears. (Quite analogously the absorption threshold of very thin CdS films deposited on the substrate made of another semiconductor gets shifted [151].) Concurrently with the increase in E_g the Fermi level also shifts. This shift offers, at least, in principle an opportunity of reducing the substances on illuminated dispersed semiconductor (of course, by using a more short-wave light), which do not get reduced on macroscopic electrodes in photoelectrochemical cells.

5.3 Photosplitting of Water

In recent years the splitting of water into hydrogen and oxygen upon illumination of semiconductor microheterogeneous systems (suspensions and colloids) has been considered as an alternative to photoelectrolysis in the cells with macroscopic electrodes described in Chap. 3. Oxygen photoevolution from aqueous suspensions of TiO_2, ZnO, and WO_3 was first noticed by Krasnovsky et al. [152]. More exactly, this was not the decomposition proper of water, because hydrogen evolution was replaced by the reduction of "sacrificial" acceptor (Fe^{3+}, $Fe(CN)_6^{3-}$, or p-benzoquinone). Therefore, free energy less than 1.23 eV per electron was stored in the system. Later on, complete decomposition of water, in the absence of sensitizers and mediators, was carried out by illuminating the platinized TiO_2 suspension with ultraviolet light in 1 N H_2SO_4 solution [153]. Afterwards, the evolution of hydrogen and oxygen was observed time and again upon illumination of the wide-gap semiconductor ($SrTiO_3$, $KTaO_3$, TiO_2) suspension whose particles carried the islets of platinum metals – catalysts (the description of particular systems and a review of the literature are available in Chaps. 7 through 10 of Ref. [2]). On the whole, their action can be explained by using the conceptions of microgalvanic pairs (see Sect. 5.1) in which n-type semiconductor acts as local anode on which water is oxidized to oxygen, and the platinum metal, as cathode where water is reduced to hydrogen.

The report of Grätzel [154] on experimental realization of the entire water-splitting scheme using a "bifunctional catalyst", shown in Fig. 67 c, attracted widespread attention. Titanium dioxide (anatase) was used: in the early experiments it was doped with niobium (to make it conductive); in the subsequent experiments, TiO_2 was also doped with chromium to make it sensitive to visible light (cf. Fig. 57). Catalysts – Pt and RuO_2 – in amounts of the order of 1 mass percent were introduced into the dispersion. The evolved gases were detected chromatographically.

In the experiments conducted with TiO_2 dispersions, the following salient features of the water-splitting process were highlighted.
a) Hydrogen and oxygen are produced in the 2:1 stoichiometric ratio.
b) In the beginning the hydrogen evolution rate remains constant, but with the passage of time it decreases. This happens due to the occurrence of a competing cathode reaction, i.e., electrolytic reduction of the formed oxygen. The relative rate of this reaction increases as oxygen is accumulated in the solution; after an inert gas is blown through the solution and oxygen is removed, the hydrogen formation rate regains its initial value.
c) After the light is switched off, the concentrations of oxygen and hydrogen in the solution do not remain constant, but start decreasing due to their recombination into water – a process catalyzed by platinum on the TiO_2 surface.

d) Though hydrogen appears in the solution practically immediately after the light is switched on, oxygen is first detected after a rather long induction period; this is because oxygen is adsorbed on the TiO_2 surface and starts coming into the solution only after the dispersion surface gets adsorption-saturated.

Also it has been found that the addition of only Pt or only RuO_2 into the TiO_2 dispersion imparts only moderate electrolytic activity to this dispersion. But when both these catalysts are added jointly, the water photosplitting rate increases by an order. The light energy conversion efficiency is yet small: the maximum reported value of this efficiency is but 0.7 percent [155].

Afterwards, in these experiments TiO_2 was replaced by CdS - as semiconductor sensitive to visible light [156]. Though a "macroscopic" CdS photoanode undergoes corrosion upon illumination, the RuO_2 microdeposits on the surface of CdS particles, as affirmed, safeguard the semiconductor against decomposition. This is because the water oxidation reaction competing with anodic decomposition of CdS for the photogenerated holes is accelerated by the catalyst to an extent that the relative rate of photocorrosion becomes negligible.

Later on, however, the correctness of interpretation of data reported in Refs. [154 and 156] was questioned.

First of all, as mentioned in Ref. [46], no one has succeeded in reproducing the results of Ref. [156]. Also, the attempts made to carry out the same process not in suspensions but in more sophisticated model systems did not yield satisfactory results. One of the model systems was a macroscopic cell with a CdS photoanode coated with a 10-500 nm thick layer of RuO_2 (sprayed on CdS in the microwave plasma). It has been revealed that this layer does not completely safeguard CdS against corrosion: in the presence of dissolved oxygen the following reaction proceeds under illumination:

$$CdS + 2O_2 \rightarrow Cd^{2+} + SO_4^{2-} \tag{5.5}$$

And hydrogen is not evolved on the cathode of a short-circuited cell [157, 158].

In another model system [159], CdS grains of size 35-45 µm were implanted in a 20-25 µm-thick polymer (polyurethane) membrane separating the aqueous electrolyte solutions in the cell, as shown in Fig. 73. The catalysts - RuO_2 and Pt - were put on the CdS surface from different sides of the membrane. No photodecomposition of water was found to take place in such a system. Yet this process

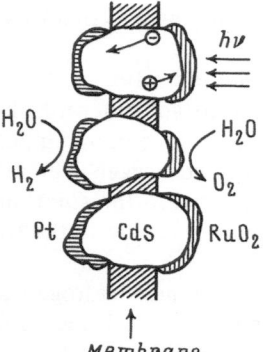

Membrane

Fig. 73. Scheme of model experiment on photosplitting of water using RuO_2/CdS/Pt system in polymer membrane [46]

occurs without any difficulty when both the catalysts are put on CdS from the same side of the membrane. This may be expounded as evidence of direct (not via free carriers in CdS) interaction of both the catalysts; this interaction may be associated with some synergistic effect whose nature is yet not known.

And, lastly, the initial conception that in a semiconductor dispersion with a "bifunctional catalyst", RuO_2 catalyzes only the evolution of O_2 from water, and Pt, only that of H_2, is likely to be oversimplified. In fact, RuO_2 is a good catalyst not only for the oxidation but also for the reduction reaction, for example, for H_2 evolution. It is quite probable that the catalyst's action depends not only and not so much on its chemical nature as on the electrical properties of its contact with the semiconductor. Thus, direct measurements [64] have shown that the CdS/Pt contact is not ohmic but is rectifying both prior to and after the absorption of hydrogen by platinum (in contrast with the TiO_2/Pt and $SrTiO_3$/Pt contacts): the barrier height ranges between 1 and 1.6 eV. Therefore, platinum on CdS can hardly be a good catalyst for the hydrogen-evolution process involving the participation of electrons (i.e., the majority carriers in CdS).

Unlike CdS, at the contact with platinum, the barrier height in TiO_2 changes significantly upon platinum hydrogenation. Therefore, the platinum clusters on TiO_2 grains can catalyze the evolution of either hydrogen (if platinum is hydrogen-charged and forms an ohmic contact with TiO_2) or oxygen (if platinum is not hydrogen-charged and the contact remains rectifying). The distribution of clusters between these two functions has a random nature; and the electrochemical process "automatically" keeps the catalyst either in the hydrogen-charged or in the hydrogen-deprived state.

The principle difficulty in the explanation of the action of a "bifunctional catalyst" given in Ref. [139] resides in the following [160]. As the catalyst accelerates a certain reaction, it should be in equilibrium with the corresponding redox system in the solution. Hence, their electrochemical potential levels should coincide. Then in the Pt/TiO_2/RuO_2 suspension for splitting of water, the catalysts should have been in equilibrium with two different systems (whose electrochemical potentials differ by about 1.23 eV) such that $F_{Pt} = F_{H_2/H_2O}$ and $F_{RuO_2} = F_{H_2O/O_2}$. This is what exactly happens in a usual electrolyzer with catalytically active metal electrodes across which the voltage is, in fact, close to the indicated value. But in a colloidal-size particle this difference in the local values of the Fermi level must create an enormous electric field of the order of 10^5-10^6 V/cm. In the material which does not exhibit the properties of an insulator (for example, Nb-doped TiO_2, see above) this field would have caused a strong electric current to flow through the particle.

If, on the other hand, both the catalysts are in equilibrium with the same redox system, say "hydrogen", such that $F_{Pt} = F_{TiO_2} = F_{RuO_2} = F_{H_2/H_2O}$, then they catalyze H_2 evolution. However, in this case, the occurrence of a second reaction, i.e., O_2 evolution, becomes impossible because the photogenerated charges in TiO_2 which have crossed the TiO_2/catalyst boundary, can further pass into the solution only as electrons with energy equal to the catalyst Fermi energy (i.e., but again F_{H_2/H_2O}).

Hence, in the considered catalyst 1/semiconductor/catalyst 2 system (see Fig. 67 c) the main advantage of the semiconductor, i.e., the possibility of carrying out simultaneously two different reactions, is offset. (Of these, one reaction involves the participation of the majority carriers with an electrochemical potential $F_n \simeq F$, and the other, of the minority carriers with an electrochemical potential $F_p \neq F$, see p. 35.)

There are two way-outs from this contradiction. It may be assumed that though Pt and RuO_2 form sufficiently large-size clusters on the semiconductor surface (as revealed by the electron microscopic studies), virtually not they but more fine formations containing only few atoms of a noble metal act as a catalyst. The properties of an individual phase, in particular, a definite value of the Fermi level cannot be assigned to these formations; but they may be regarded as surface states. Then the minority carriers can react at an electrochemical potential equal to their quasi-Fermi level F_p, irrespective of the electrochemical potential of the reaction involving the participation of majority carriers. In other words, the minority carriers' reaction is catalyzed on the atomic but not on the "phase" level.

An alternative assumption is that the islets of RuO_2 and Pt catalysts are virtually deposited on different but not on the same particle of the semiconductor. Thus, the suspension consists of two kinds of particles: TiO_2/RuO_2 and TiO_2/Pt. (In fact, direct evidence of the presence of both types of catalysts on the surface of the same particle have not been produced.) And the process proceeds not in accord with the scheme presented in Fig. 67c, but in the following manner. On the particles containing Pt islets, a certain amount of hydrogen evolves with the transfer of electrons and the hole charge is accumulated; on the particles containing RuO_2 the evolution of oxygen takes place with the participation of holes and the electron charge is accumulated. Accumulation of charge on the particle halts the enumerated reactions. These charges neutralize upon collision of different "kinds" of particles and the process may proceed further. A simple estimate [161] reveals that this inter-particle exchange of charges is not necessarily a slow stage. And experimentally it has been proved that such a particle-to-particle transfer of electrons is possible in the system for hydrogen photoevolution. In this system the semiconductor (CdS) and the catalyst (Pt) are first separately deposited on the particles of different inert carriers (SiO_2 and other oxides) and then the two suspensions are mixed. Hydrogen production starts when such a mixture is illuminated. This implies that the electrons photogenerated in CdS readily transfer upon collision to the particles containing Pt clusters, where hydrogen evolves [162].

On the whole, the mechanism of photosplitting of water in semiconductor suspensions with "bifunctional catalysts" yet demands a detailed commentary.

5.4 Photodecomposition of H_2S and Sulfides. Other Reactions Involving the Participation of Inorganic and Organic Compounds

The reaction of decomposition of hydrogen sulfide (see Eq. (3.8)) and of sulfides is very essential for the purification of waste waters and as a source of hydrogen at the same time. The authors of Ref. [163] have succeeded in carrying out this reaction by illuminating with visible light the RuO_2-coated CdS suspension particles in aqueous solutions of sulfides [163]. This reaction proceeds at a very fast rate when the RuO_2 content in the dispersion is of the order of 1 %: the H_2-formation quantum yield reaches 35 %. Sulfur that precipitates out in the course of decomposition of sulfides practically does not slow-down the reaction. The presence of SO_3^{2-} ions in the solution makes it possible to obtain, together with H_2, another valuable product – thiosulfate – according to the reaction $S + SO_3^{2-} \rightarrow S_2O_3^{2-}$ [164]. The photodecomposition of hydrogen sulfide at the particles with a Cu_xS/CdS heterojunction, obtained by co-coagulation of colloidal solutions of these two sulfides, is effective in particular [165]. Here, the photovoltages across two interfaces – CdS/solution and Cu_xS/CdS – are added up. On the whole, such a particle, according to its action mechanism, resembles a photoelectrochemical cell with a "tandem" electrode (see Sect. 3.3.3).

Some other oxidation and reduction reactions occurring in semiconductor suspensions and colloidal solutions are briefly listed in Table 5.2. The review cited under Ref. [166] is specially dedicated to the reactions of organic substances. We shall supplement these with photoprecipitation of heavy metals (Pt, Pd, Au, Cu, and others) from the solutions of their salts and photooxidation of surface-active substances (see, for example, Ref. [167]) on semiconductor particles. All these reactions are a potential tool for freeing waste waters from harmful products and/or for obtaining chemical fuels.

Unlike the photoelectrochemical cells, spatial separation of photoelectrochemical reaction products does not take place in microheterogeneous systems. In some cases the separation of the desired product does not present specific difficulties (if the product separates out as gas or precipitates); in other cases it be-

Table 5.2. Photoelectrochemical oxidation and reduction reactions in semiconductor microheterogeneous systems

Reagents	Semiconductor	Products
Oxidation and dehydrogenation reactions		
CN^-	TiO_2, CdS, ZnO	OCN^-
Alcohols (in the atmosphere of O_2, N_2O)	TiO_2	Ketones
Benzene, toluene	TiO_2	Hydroxylation products
Aromatic amines	ZnO	Azo compounds
Cl^-, Br^-, I^-	TiO_2/Pt	Cl_2, Br_2, I_2
CO	TiO_2/Pt	CO_2
C_2H_5OH	TiO_2/Pt, TiO_2/RuO_2	$CH_4 + H_2$
CH_3COOH	TiO_2/Pt, TiO_2/RuO_2	CH_4, $C_2H_6 + CO_2$ [a]
Glucose and other carbohydrates	TiO_2/Pt, TiO_2/RuO_2	$H_2 + CO_2$
Reduction reactions		
CO_2	$SrTiO_3$, WO_3, TiO_2, ZnO, Fe_2O_3, and others	$HCHO + CH_3OH + CH_4$
N_2	TiO_2	NH_3, N_2H_4

[a] The so-called "photo-Kolbe reaction", see Eq. (4.2).

comes a problem (for instance, separation of the water decomposition products, H_2 and O_2). In some particular cases this problem can be attacked by implanting the semiconductor microparticles within the membrane in the manner as shown in Fig. 73. Such a device enables the photoelectrolysis products to be spatially separated by maintaining the main advantage of microheterogeneous systems, i.e., high specific surface of the semiconductor.

Thus, in a sulfonated polyperfluoroethylene-based membrane[2], the thickness being of the order of 0.1 mm, the CdS, CdS + ZnS, and TiO_2 particles are built-in by impregnating the membrane with the corresponding metal salts and treating it with hydrogen sulfide or an oxidizing agent. Upon illumination of these membranes the same reactions as in the suspensions occur, e.g., photooxidation of Br^- to Br_2 (on TiO_2) or H_2 evolution from water with the use of a "sacrificial" donor (on CdS) [168].

Another type of membrane – bilayer lipid – was used in Ref. [169] for creating not a "macroscopic" membrane system described above and shown schematically in Fig. 73, but again a microheterogeneous, namely a vesicular, system. The vesicles are closed formations whose wall represents a membrane; the inside and outside liquid may have different composition. If semiconductor particles are implanted in the vesicle wall, then, upon illumination, photochemical reactions occur and the composition of the medium within the vesicles and around them changes. On completion of the process, the formed products are removed by one or another method from the inner and outer space.

Though the studies of microheterogeneous systems have not yet led to the creation of high-efficiency solar energy converters that could compete with the electrochemical cells for photoelectrolysis, described in Chaps. 3 and 4, or for obtaining electrical current (see Chap. 6), yet the potential possibilities of this trend are great beyond doubt.

---

[2] This solid polymer electrolyte with proton conductivity is known as Nafion. It is used in the electrolyzers described in Sect. 3.4.1.

Chapter 6
Solar Energy Conversion into Electrical Energy. Regenerative Photoelectrochemical Cells

6.1 Basic Problems

Regenerative photoelectrochemical cells whose operation principle is described in Sect. 2.1 are, as their other name – "liquid-junction solar cells" – implies, electrochemical analogs of solid-state solar cells. The latter are available in two variants (see, for instance, Ref. [170]): (1) with a p-n-junction in the semiconductor, and (2) with a semiconductor/metal contact (the so-called Schottky diodes). The photoelectrochemical cells are called upon to replace just this second variant.

The potential benefits (as well as the possible drawbacks) of this replacement can be conveniently discussed by referring to the cell operation scheme. If in Fig. 21 the middle part (electrolyte solution) is "removed" and the semiconductor and metal phases are moved together until they touch each other, then one obtains the operation scheme of an illuminated solid-state Schottky diode. Here, the semiconductor is illuminated through the metal layer. The generation and transportation of carriers in a semiconductor take place similarly in both types of cells (electrochemical or solid-state). However, in solid-state Schottky diodes the light-generated minority carriers are transferred from the semiconductor directly into the metal (as this takes place, the holes, cf. Fig. 21, recombine with the metal electrons at the semiconductor/metal interface), while in photoelectrochemical cells the transfer of charges from the semiconductor into the metal includes, as an intermediate stage, two electrochemical reactions (direct and reverse) occurring on the cell electrodes.

What can be the consequences of the replacement of a direct semiconductor-to-metal contact by a three-phase semiconductor/electrolyte solution/metal system? First of all, now it is not necessary to illuminate the semiconductor through the metal layer. The metal layer, however thin it may be, inevitably absorbs part of the light energy which is lost never to return. The decrease in film thickness, however, entails an increase in its electrical resistance; the so-called lateral resistance manifests itself in the electric circuit of the cell, and this impairs the current-voltage characteristic. Illumination of the semiconductor through the electrolyte provides, at least, in principle a means for eliminating these interferences (if the electrolyte is sufficiently conductive and transparent).

Another circumstance is associated with the nature proper of the interface. At the contact of two solid phases (semiconductor and metal) mechanical stresses appear due to the mismatching of crystal lattices, which entail electric losses (due to contact resistances, surface states, recombination, etc.). The semiconductor surface in contact with the liquid phase (electrolyte solution) remains "perfect"; it does not experience mechanical stresses. The density of surface states is therefore small (or it may be lowered by giving certain chemical treatments). All this aids in obtaining a higher photoelectrochemical energy conversion efficiency.

Of course, the employment of a liquid contact has its drawbacks too. The chemical action of a solution on the semiconductor may produce undesirable effects such as corrosion, ion exchange (see Sect. 6.2.1) and others. Besides, the charge transfer at the electrode/solution interface does not always proceed at a sufficiently high rate. Thus, the task resides in choosing such particular systems in which the advantages of the electrochemical approach predominate over its disadvantages.

The efficiency of liquid-junction (as well as solid-state) solar cells can, according to Eq. (2.11), be expressed as:

$$\eta = \frac{i_{sh.c} \cdot \varphi_{ph}^{o.c} \cdot f}{P_l} \tag{6.1}$$

Here, $i_{sh.c}$ is the short-circuit current, $\varphi_{ph}^{o.c}$ the open-circuit photovoltage, and f the fill factor of the cell; P_l is the power density of the incident light flux.

Using Eq. (2.16) for the fill factor, one can rewrite Eq. (6.1) in the form:

$$\eta = \frac{i_{ph}^{MPP} \cdot \varphi_{ph}^{MPP}}{P_l} \tag{6.2}$$

where i_{ph}^{MPP} and φ_{ph}^{MPP} are respectively the photocurrent and photovoltage at the maximum power point (MPP) (cf. Fig. 28).

For increasing the value of quantities appearing in the numerator of Eq. (6.1) or (6.2) (and at the same time the efficiency) it is necessary first of all to decrease the overvoltage at the cell electrodes (i.e., to have a highly reversible redox system in the solution). This follows from the energy balance of the energy conversion process, which, by analogy with Eq. (3.2) for photoelectrolysis, can be expressed as:

$$F_n - F_p = e\,(i_{ph}^{MPP} \cdot R + \eta_a + \eta_c) \tag{6.3}$$

where F_n and F_p are respectively the quasi-Fermi levels of electrons and holes in an illuminated semiconductor. (The difference in these levels depends on the light intensity J_0, light absorption coefficient, recombination rate, and other factors.) η_a and η_c are overvoltages, respectively, at the anode and cathode of the cell,[1] R is the load resistance. (More exactly, R includes also the internal resistance of the cell, which must be reduced as far as possible so as not to decrease the fill factor; see below.) The voltage tapped off from this load equals $i_{ph}R$, and the power, i_{ph}^2R. The power delivered by a solar cell depends on the light intensity and the load resistance. The latter is so chosen that the solar cell would yield maximum power.

---

[1] Here, for a semiconductor electrode, one considers only that part of total overvoltage which is localized in the Helmholtz layer (i.e., η_H).

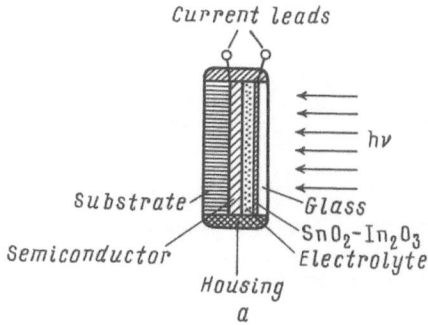

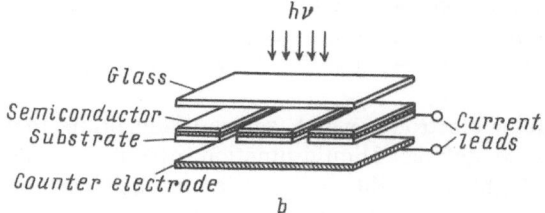

Fig. 74. Thin-layer design of liquid-junction solar cell (a) optical window is combined with counter-electrode; (b) optical window and counter-electrode are placed on opposite sides of the photoelectrode

A thin-layer plane parallel cell design is used to reduce electric losses and to enhance, at the same time, the mass transfer between the cell electrodes. In addition, such a design assures small losses by light absorption in solution. One of its variants is shown in Fig. 74a. The semiconductor electrode is illuminated through a glass window whose inner surface is covered with a conductive transparent layer, e.g., of SnO_2-In_2O_3 which acts as a counter-electrode. The thickness of the layer of electrolyte between the window and the photoelectrode usually equals about one tenth of a millimeter.

Another variant of a thin-layer design of a photoelectrochemical cell is represented in Fig. 74b. Here, the counter-electrode is placed behind the photoelectrode. The latter consists of separate sections between which openings are provided for the current to flow from the frontal (illuminated) photoelectrode surface to the counter-electrode. For reducing ohmic losses and obtaining maximum possible efficiency, the geometric dimensions of the cell must be optimized. Such an optimization calculation has been performed in Ref. [171]. Given a certain specific resistance of the electrolyte, the authors of Ref. [171] computed the dependence of the delivered electric power P_{el} on the dimensionless parameters L/h and h/G, where L is the half-width of the photoelectrode section, G is the half-width of the openings for the flow of current between the sections, and h is the distance between the cell window and the illuminated photoelectrode surface, and between the photoelectrode rear surface and the counter-electrode. As evident from Fig. 75, in this design the solution layer thickness, h, between the photoelectrode and the window, on the one hand, and the counter-electrode, on the other hand, should be rather large to attain maximum efficiency. This is a drawback, compared to the type of the cell shown in Fig. 74a, because the thicker the solution layer, the more

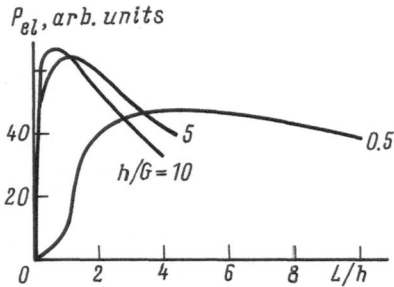

P_{el}, arb. units

40

20

$h/G = 10$

5

0.5

0   2   4   6   8 L/h

Fig. 75. Computed dependence of power produced by a liquid-junction solar cell (Fig. 74 b) on its geometrical dimensions (see text) [171]

the light absorbed in it. In Ref. [171] a calculation has also been performed for a cell in which a gauze-type counter-electrode is placed between the optical window and the photoelectrode, thus partially shadowing the latter. Because of shadowing, maximum efficiency turned out to be even less than for the variant illustrated in Fig. 74 b.

Photocorrosion of semiconductor electrodes which substantially limits their service life is the main obstacle to the making of liquid-junction solar cells. Photocorrosion is suppressed by using a highly reversible redox system in the electrolyte solution so that, for example, the oxidation-reaction of the reduced component successfully competes for light-generated holes with the electrode material anodic photodecomposition reaction (cf. Sect. 2.2). This very redox system serves also to carry current between the photoelectrode and the counter-electrode. Along with aqueous solutions, non-aqueous solutions have found wide use in recent years. In the latter, the semiconductor materials are less susceptible to corrosion. Both inorganic and organic compounds are used as effective redox systems, in particular, ferrocene and other metalocenes [172].

This method of protecting semiconductor electrodes permits the use of materials with optimum bandgap width – GaAs, InP, CdTe, MoSe$_2$, Si, and others – as photoanodes, and allows to obtain conversion efficiencies for liquid-junction solar cells close to those for solid-state solar cells.

It may be noted, however, that even the potential danger of corrosion places limitations on the efficiency of regenerative photoelectrochemical cells. In fact, as already mentioned in Sect. 2.1, the initial (i.e., "dark") band bending in a semiconductor must be increased to raise the photovoltage (and, hence, the efficiency). For this purpose, the reversible potential φ^0 of the redox system that sets band bending, say, in a photoanode must be as positive as possible. At the same time, however, it should not exceed the electrode material's anodic decomposition potential $\varphi^0_{dec,p}$. This compells to deliberately reduce the photopotential of the cell. Thus, as the solution's oxidative ability increases, the efficiency grows and the stability decreases. And the real characteristics of photoelectrochemical cells are established as an outcome of the compromise between the effectivity and reliability requirements.

It is just the need to combat photocorrosion than can be attributed to the increased interest in recent years in photocathodes, as they are less susceptible to corrosion than photoanodes. (This equally holds both for the regenerative cells and the cells for photoelectrolysis; cf. Sect. 3.2.)

Let us summarize the enumerated basic requirements on the components of a photoelectrochemical cell for converting solar energy into electrical energy. The requirements that must be placed on the semiconductor photoelectrode material are formulated in Sect. 2.4 (they are in essence common to liquid-junction solar cells and the cells for photoelectrolysis). The following requirements are placed on the electrolyte solution.

1. The redox system in the solution should have a reversible potential φ^0 such that the condition $\varphi^0 > \varphi_{fb}$ (for a n-type semiconductor electrode) or $\varphi^0 < \varphi_{fb}$ (for a p-type semiconductor electrode) holds. The difference $|\varphi^0 - \varphi_{fb}|$ must be as much as possible (on the condition that requirement (2) is fulfilled).
2. The condition $\varphi^0 < \varphi_{dec,p}^0$ (for a n-type semiconductor electrode) or $\varphi^0 > \varphi_{dec,n}^0$ (for a p-type semiconductor electrode) must be complied with.
3. The electrochemical reactions on the cell electrodes should be highly reversible, in other words, the overvoltage η_a (or η_c) should be low.
4. The specific resistance should be small.
5. The light transmission in the visible range of spectrum should be rather large.

Table 6.1 lists the characteristics of some of the most effective regenerative photoelectrochemical cells described to date. From the data of this table an important conclusion can be made, in particular: polycrystalline electrodes exhibit high (both absolute and relative as compared to monocrystalline electrodes) efficiencies. This enables one to look forward to the creation of cheap and simple-to-manufacture solar energy converters.

Though it is a bit premature to talk about the economic aspects of the application of liquid-junction solar cells, we shall nevertheless give the results of the estimates of admissible capital outlays on making cells, at which the solar energy con-

Table 6.1. Efficiency of regenerative photoelectrochemical cells

Photoelectrochemical cell	Efficiency [a], %	Ref.
n-GaAs (monocrystalline) $\vert$ K_2Se–K_2Se_2–KOH $\vert$ C	12	[173]
n-GaAs (polycrystalline) $\vert$ K_2Se–K_2Se_2–KOH $\vert$ C	7.8	[173]
n-GaAs (monocrystalline) $\vert$ K_2Se–K_2Se_2–KOH $\vert$ C	15	[174]
n-GaAs$_{1-x}$P$_x$ (epitaxial) $\vert$ Ferrocene-ferricenium $\vert$ Pt (acetonitrile solution)	13.2	[175]
n-CdSe (monocrystalline) $\vert$ Na_2S–Na_2S_2–NaOH $\vert$ C	7.5	[176]
n-CdSe (polycrystalline) $\vert$ Na_2S–Na_2S_2–NaOH $\vert$ C	5.1	[176]
p-InP (monocrystalline) $\vert$ VCl_3–VCl_2–HCl $\vert$ C	11.5	[173]
p-InP (polycrystalline) $\vert$ VCl_3–VCl_2–HCl $\vert$ C	7	[177]
n-Si (monocrystalline) $\vert$ HBr–Br_2 $\vert$ Pt	11.4	[65]
n-Si (monocrystalline) $\vert$ Ferrocene-ferricenium $\vert$ Pt (methanolic solution)	10.1	[178]
n-Si (amorphous) $\vert$ Ferrocene-ferricenium $\vert$ Pt (methanolic solution)	3	[179]
n-WSe$_2$ (monocrystalline) $\vert$ KI_3–KI $\vert$ Pt	10.2	[180]
n-CuInSe$_2$ (monocrystalline) $\vert$ KI_3–KI–HI $\vert$ Pt	12	[181]
n-CuInSe$_2$ (polycrystalline) $\vert$ KI_3–KI–CuI–$In_2(SO_4)_3$–HI $\vert$ Pt	12	[182]

[a] Sunlight (natural or simulated).

version can become competitive. Calculations have been performed in Refs. [171] and [183] for the accepted margins of profit in the USA, with a 5-year payback period, and a power density of incident light of 25 mW/cm² when averaged for 24 hours. The authors of these works have found that for an efficiency of 8 to 15 %, the upper limit of capital outlays ranges between 40 and 85 US-dollars per m² of cell area. Note that this is cheaper by one order compared to the 1983 prices of silicon solid-state solar cells mentioned in Sect. 3.4.2. Thus, primary emphasis must be given to the development of polycrystalline electrodes and simple and cheap designs of photoelectrochemical cells.

6.2 Description of Particular Systems

6.2.1 Cadmium Chalcogenide Electrodes

One of the main classes of regenerative photoelectrochemical cells has been constructed by using n-type cadmium chalcogenide photoanodes in solutions containing chalcogenide/polychalcogenide redox system, i.e., $CdX \,|\, X^{2-}-X_n^{2-}$, where X = S, Se, Te (and n is generally ≥ 2). The key advantage of these semiconductor materials resides in preparing relatively efficient photoelectrodes by simple and cheap methods. In principle, these methods are appropriate for obtaining large-area electrodes with standardized properties.

The results of the first phase of studies made into these cells are listed in Sect. 8.4 of Ref. [1]; the results of more recent studies are reviewed in Ref. [184]. The photoelectrochemistry of cadmium chalcogenide/polychalcogenide electrolyte systems is considered in detail in the review by Hodes (Chap. 13 of Ref. [2]). Without going into the details of physicochemical analysis of the numerous studied systems, below we shall summarize the state-of-the-art of photoelectrochemical cells.

Electrodes. Cadmium sulfide with its too wide forbidden band ($E_g = 2.4$ eV) is less suitable for conversion just of solar radiation (cf. Fig. 27). The forbidden bandwidth in cadmium selenide (1.74 eV) and cadmium telluride (1.5 eV) is close to the optimum one. Still better is the situation in solid solutions of type $CdSe_xTe_{1-x}$; at $x \simeq 0.5$ the value of E_g passes through a minimum (about 1.4 eV), Fig. 76 a. In this very range of compositions, the $CdSe_xTe_{1-x}$ electrode exhibits also increased stability when used in a polysulfide electrolyte (Fig. 76 b).

Polycrystalline samples[2] are prepared by a number of methods [184].
1. Preparation of ceramic electrodes:
    a) by pressing the powders into tablets and their subsequent sintering (for example, CdS, CdSe, Cd(S, Se), CdTe, WSe_2, $MoSe_2$);
    b) by smearing the semiconductor powder/binder pulp over the substrate, followed by heating (Cd(Se, Te), MoS_2, $CuInS_2$);

---

[2] The examples also include some other materials besides CdX.

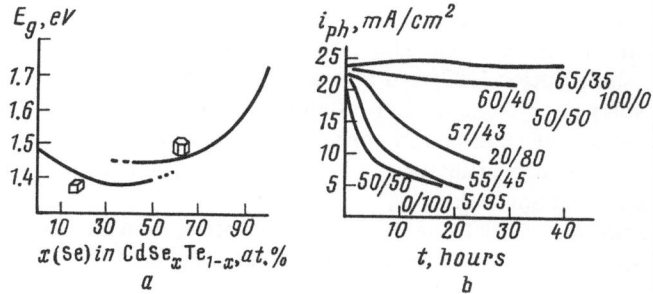

Fig. 76. Dependence of forbidden bandwidth of $CdSe_xTe_{1-x}$
(a) and its stability upon operating in $2\,M\,KOH + 2\,M\,Na_2S + 2\,M\,S$ solution (b) on concentration of Se and on crystal structure of material [243]
Cubic structure is typical of pure CdTe; hexagonal structure is typical of CdSe; both phases may coexist in the intermediate ranges of composition. Top figures on curves give ratios of concentrations Se:Te, and lower, the ratio of contents of phases with hexagonal and cubic lattices (annealed paste samples) (Reprinted with permission from Journal of American Chemical Society. Copyright (1980) American Chemical Society)

2. Vacuum deposition methods:
   a) vacuum evaporation (CdS, CdSe, Cd(Se, Te));
   b) glow discharge precipitation (a-Si:H);
3. Electrolytic deposition
   a) cathodic electrodeposition (see also the review cited under Ref. [185]) (CdSe, CdTe);
   b) anodizing the metal substrate in chalcogenide solutions (CdS, Bi_2S_3);
4. Chemical methods:
   a) chemical vapor deposition (CVD) (GaAs, InP)
   b) pyrolysis of salts solution by spraying them onto the heated substrate (CdS, CdSe);
   c) electroless deposition (CdS, CdSe, Bi_2S_3).

Titanium is more commonly used as substrate for photoanodes, because if the deposited semiconductor-layer is porous, the portions of the substrate surface accessible for the electrolyte deep in the pores readily get clogged with anodic oxide (TiO_2) which blocks further the flow of current through the pores. Use is also made of chromium and graphite substrates. The details of production procedures and techniques are available in some of the below-cited works (see also the literature in Ref. [184] and Chap. 13 in Ref. [2]).

A very significant step of the procedure of obtaining highly photosensitive electrodes is photoelectrochemical etching of the semiconductor, which often follows chemical etching. It is carried out upon anodic polarization in the etching bath (for example, for CdSe, in an $HNO_3-HCl-H_2O$ (0.3:9.7:90) mixture [186]). Thereby the defects, fortuitous impurities, and surface states usually present on the surface and in the near-surface layer of the as-machined semiconductor sample are removed. With this treatment the photosensitivity and the stability of photoelectrodes increase; in addition, their photoelectrochemical, photoluminescence, and other characteristics improve (for instance, the quantum yield of photocurrent on CdSe reaches 0.85 [186]).

In a number of cases an additional gain in the photosensitivity of electrodes is attained after ion adsorption on the surface. Thus, treating the surface of CdS and Cd(Se, Te) with a copper salt solution [187] and also of CdSe with a zinc salt solution [188] increases the solar energy conversion efficiency. In most cases the detailed mechanism of the action of etching and surface modification on the physicochemical and electrophysical properties of the surface is not yet known, both these treatments being merely empirical.

Attempts had been made to use chalcogenides of metals other than Cd, for instance, ZnS, Bi_2S_3, and also $CdInSe_4$, but without much success (see Chap. 13 in Ref. [2]).

Electrolyte solutions. As a rule, chalcogenide-polychalcogenide redox systems protect cadmium chalcogenides well against photocorrosion in accordance with the principles formulated in Sect. 2.2. Indeed, it is the thermodynamic causes that play a leading role in stabilizing, say, a CdS photoelectrode. This is evident from Fig. 77 where the dependence of the measured "stabilization factor" (i.e., the ratio of photooxidation current of the reductant (Red) to the sum of this current and the CdS photoanodic dissolution current) on the system's reversible potential φ^0 is presented. Indeed, as it follows from the quasi-thermodynamic approach, the more negative the φ^0, the higher the degree of stabilization; e.g., in SO_3^{2-} and S^{2-} solutions practically the entire anodic photocurrent is spent to oxidize the solution and therefore photocorrosion does not proceed. (But this, of course, does not discard the consideration of kinetic factors that may limit the cadmium chalcogenide corrosion rate.) A summary of computed decomposition potentials of cadmium (as well as zinc) chalcogenides, given in Chap. 13 of Ref. [2], enables one to evaluate the photoelectrochemical stability of the discussed systems.

The protective action of the chalcogenide electrolyte enhances with increase in its concentration. Usually, alkaline solutions of almost equal concentrations (1 mol/l or more) of X^{2-}, X_2^{2-}, and OH^- ions are used. In the cells with CdS electrodes, best results (on stability and efficiency) are obtained by using the $Se^{2-}-Se_2^{2-}$ system, and with CdSe and Cd(Se, Te) electrodes, by using the $S^{2-}-S_2^{2-}$ system.

On increasing the illumination intensity (and, hence, the photocurrent), the stability of chalcogenide electrodes in the chalcogenide ion electrolytes decreases. (This decrease may be explained within the scope of kinetic models of photocorrosion, enumerated in Sect. 2.2; see also Ref. [186].) Figure 78 shows that the CdSe

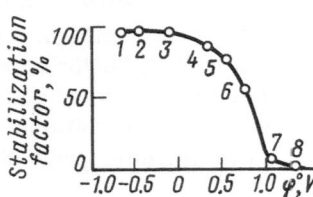

Fig. 77. Comparative efficiency of suppressing photoanodic dissolution of CdS by adding different reductants (in amounts of 0.01 mol/l) to the background solution (0.2 M Na_2SO_4) depending on their reversible redox potential [189] Y axis - stabilization factor determined by the method of rotating ring-disc electrode
1 - SO_3^{2-}; 2 - S^{2-}; 3 - $S_2O_3^{2-}$; 4 - $Fe(CN)_6^{4-}$; 5 - I^-; 6 - Fe^{2+}; 7 - Br^-; 8 - Cl^- (Reprinted by the permission of the publisher, The Electrochemical Society, Inc.)

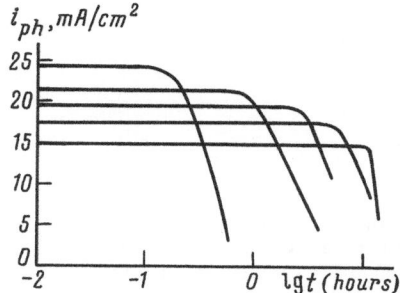

$i_{ph}, mA/cm^2$

Fig. 78. Effect of illumination intensity on the stability of monocrystalline CdSe photoanode in $1\,M$KOH + $1\,M$Na$_2$S + $1\,M$S solution [244] Initial value of photocurrent serves as measure of illumination intensity (Reprinted by the permission of the publisher, The Electrochemical Society, Inc.)

photoanode goes out of service (this manifests itself by an abrupt decrease in photocurrent) much faster if the illumination intensity is increased. To compensate for the effect of high light intensity, one has to raise the chalcogenide ion concentration in the solution.

Polychalcogenide electrolytes normally permit of sufficiently large initial band bending in a semiconductor (due to large difference $|\varphi^0 - \varphi_{fb}|$) and, hence, large photopotentials. Specific adsorption of S^{2-} anions makes the flat band potential more negative, as a result of which the difference $|\varphi^0 - \varphi_{fb}|$ and the photopotential $|\varphi_{ph}|$ increase still further.

As a rule, a chalcogenide-polychalcogenide ions system is rather highly reversible at a chalcogenide photoelectrode; therefore, the reaction rate at the electrochemical stage proper does not generally limit the cell characteristics. As to counter-electrode, catalytically active carbon cathodes containing teflon as a binder and cobalt and nickel as catalysts, and also copper sulfide and cobalt sulfide cathodes [190] were specially developed for liquid-junction solar cells with polychalcogenide electrolytes. On such electrodes the polychalcogenide ions cathodic reduction overvoltage does not exceed 25 mV at a current density of 10 mA/cm² (typical for photoelectrochemical cells utilizing not concentrated sunlight).

High energy conversion efficiency can be obtained by properly selecting, besides the composition and concentration of the chalcogenide-polychalcogenide redox system (i.e., of anions), the ionic composition of the electrolyte as a whole. In Ref. [191] it has been shown that the cation nature is fairly critical for solar cell operation. For example, the photocurrent of the CdSe electrode in $1\,M\,S^{2-}$ $+ 1\,M\,S + 1\,M\,OH^-$ solution containing different cations increase in the following order: Li < Na < K < Cs; in the same order improves the shape of the photocurrent-voltage curve (Fig. 79). In a photoelectrochemical cell the replacement of Li$^+$ by Cs$^+$ causes all its characteristics (open-circuit photovoltage, short-circuit photocurrent, fill factor) to improve; the cell efficiency increases by 1.7 times and the stability improves. Apparently, this happens due to both the acceleration of electrochemical reactions on electrodes and the improvement in the bulk characteristics (first of all, conductivity) of the electrolyte. The choice of alkali concentration is of major importance. For example, according to Ref. [192], the cell with a CdSe$_{0.65}$Te$_{0.35}$ electrode in a $1.8\,M\,K_2S + 3\,M\,S$ solution functions better in the absence of alkali; the addition of KOH but impairs its characteristics.

Besides the above-enumerated advantages, polychalcogenide electrolytes have

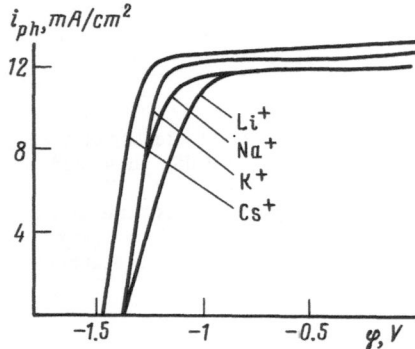

$i_{ph}, mA/cm^2$

Li^+
Na^+
K^+
Cs^+

φ, V

Fig. 79. Anodic current-voltage curves on monocrystalline CdSe photoelectrode in $1\,M\,S^{2-} + 1\,M\,S + 1\,M\,OH^-$ solution for different cations [191] (Reprinted by the permission of the publisher, The Electrochemical Society, Inc.)

disadvantages too which in any case must be allowed for in choosing the photoelectrochemical cell design. Polyselenide and polytelluride compounds are toxic and are readily oxidized by the oxygen of air. Therefore, they are suitable for use only in hermetically sealed cells. All polychalcogenide ion solutions absorb visible light [193]: polysulfide solution absorbs light of wavelength over 500 nm, and polytelluride solution, in the range between 450 and 650 nm (absorption being maximum at 512 nm). That is why the solution layer thickness through which light goes to the photoelectrode should be as small as possible.

One of the major causes of deterioration of the properties of CdS and CdSe photoanodes when used in chalcogenide electrolytes is that ion exchange takes place in the near-surface layer of the electrode. Generally speaking, in a photoelectrochemical cell the CdX electrode can be used in any combination with the redox system X^{2-}-X_n^{2-}, where X = S, Se, Te (excepting the $CdTe/S^{2-}$-S_2^{2-} system which is unstable [193]). But best results are obtained when the chalcogens in the electrode and in the solution have different nature: for example, for CdSe and Cd(Se, Te) electrodes in S^{2-}-S_2^{2-} solution (and model studies have been made, as already mentioned, in the CdS/Se^{2-}-Se_2^{2-} system). Under these conditions, in the near-surface layer of the semiconductor, the chalcogen in the crystal lattice is replaced by that of the solution. This exchange takes place in both directions: CdS in the Se^{2-} solution changes into CdSe, and CdSe in the S^{2-} solution, into CdS. It affects a layer of several tens of nanometer in thickness (i.e., hundreds of atomic layers). The exchange is particularly intense under illumination, when a photocurrent flows; and it occurs even in the dark, though at a much lower rate. The so formed restructured layer of CdS or CdSe has a very fine-grained structure.

As a result of this exchange the photocurrent decreases. But what is the particular cause of this decrease? To know this, we consider the energy diagram of the CdS electrode covered with a layer of CdSe, and of the CdSe electrode with a CdS layer on its surface (Fig. 80). Owing to the difference in the width of forbidden bands, a potential barrier may appear at the CdS/CdSe heterojunction which hinders the flow of light-generated current carriers. This has already been approached in Sect. 3.3.3 in the context of the operation of a heterojunction photoelectrode (cf. Figs. 80 and 43). As seen from Fig. 80b, the outer CdSe-layer does not hinder the hole transfer from the CdS-electrode bulk into the solution (here, one must

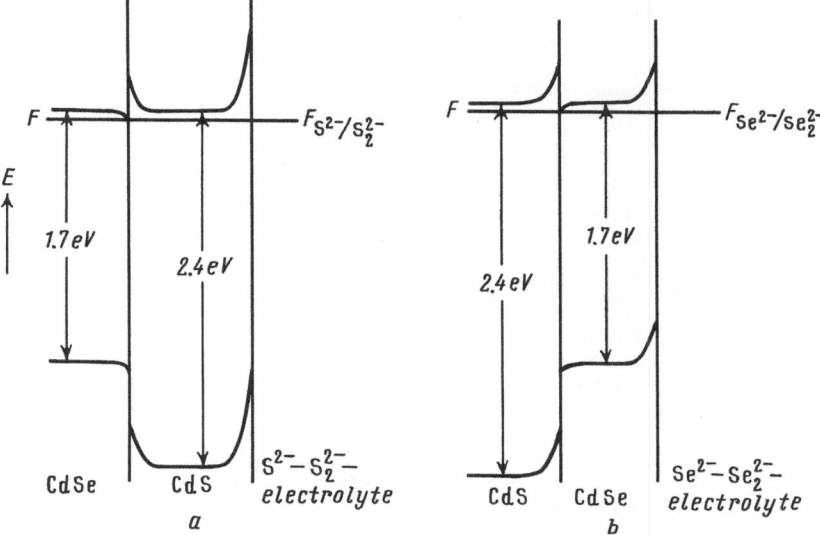

Fig. 80. Energy diagrams of cadmium chalcogenide electrodes that have been subjected to ion exchange in chalcogenide electrolytes [194]
(a) CdSe in S^{2-}-S_2^{2-} solution; (b) CdS in Se^{2-}-Se_2^{2-} solution (Reprinted by the permission of the publisher, The Electrochemical Society, Inc.)

remember that the narrow-band semiconductor layer on the wide-gap semiconductor surface attenuates light which is active for the wide-gap semiconductor). Conversely, the CdS layer on the CdSe surface blocks the transfer of in-depth holes so that only the holes light-generated in the thin outer layer participate in the photoelectrochemical reaction. (Of course, the free carriers can tunnel through the outer layer if its thickness is small (1–2 nm) and the photoactivity of the electrode is not impaired.)

Besides the formation of a potential barrier for the minority carriers, the operation of the electrode subjected to ion exchange is impaired by the increase in the recombination velocity at the defects in the restructured layer and at the junction of two semiconductors.

It is interesting that the addition of even small amounts of Se (7.5×10^{-2} mol/l) into the sulfide solution practically inhibits the replacement of selenium in CdSe by sulfur, thus making the operation of the photoanode stable [195].

Characteristics of photoelectrochemical cells. Liquid-junction solar cells with polycrystalline electrodes have a rather high solar energy conversion efficiency, as can be seen from Table 6.2. In the majority of papers, the service life of these cells has not been determined; an idea about their service life can be gathered only from rare reports. A solar cell consisting of 64 separate modules with CdSe electrodes, each having an area of 3 cm², and a total efficiency of 2.5 %, was tested under field conditions. It worked for more than 5 months; it stopped functioning due

Table 6.2. Characteristics of regenerative photoelectrochemical cells with cadmium chalcogenide polycrystalline electrodes in polysulfide electrolytes under solar illumination

Electrode material	Electrode manufacturing method	Electrode pretreatment	Efficiency [a], %	Ref.
CdSe	Chemical deposition from Na_2SeSO_3 + $Cd(CH_3COO)_2$ + NH_4OH solution		6.8	[196]
CdSe	Spraying salt solution on the heated Ti substrate; annealing in the air (at 450 °C for 30 min.)	Photoelectrochemical etching in 1 M NaCl (pH 2.5)	6.4	[197]
CdSe	Electrodeposition from $SeSO_3^{2-}$ solution; annealing at 600 °C	Dipping in $ZnCl_2$ solution	7.3	[198]
CdSe	Electrodeposition from LiCl-KCl melt containing admixtures of $CdCl_2$ and Na_2SeO_3		6.1	[199]
CdSe	Electrodeposition on SnO_2 substrate	Etching in aqua regia; dipping in 1 M $ZnCl_2$ solution	5.2	[188]
$CdSe_{0.8}Te_{0.2}$	Vacuum evaporation on heated (350–450 °C) Ti substrate		7.4	[200]
$CdSe_xTe_{1-x}$ ($0.5 \leqq x \leqq 0.8$)	Spraying CdSe, CdTe, and $CdCl_2$ suspensions on Ti substrate; baking at 600–650 °C for 15–30 min		4.5–5.5	[201]
$CdSe_{0.65}Te_{0.35}$	Paste electrode		4.1	[191]

[a] Natural or simulated sunlight.

to the leakage of electrolyte. The authors of Ref. [2] (see Chap. 13) believe that the prognosted life comes to several years.

An analogous cell functioned for several months in the open air in a relatively cold climate (in West Berlin where the ambient temperature sometimes lowered down to −17 to −20 °C in the winter) without significantly impairing the characteristics [202].

A separate study is called for to understand how the ambient temperature affects the operation of liquid-junction solar cells, and to control the thermal conditions of the cells functioning in open air. To date, this problem has not been studied in detail. From the general conceptions it is clear that ambient temperature has multiple effects on the operation of photoelectrochemical cells. Thus, upon heating the electrochemical reactions and mass transfer in solutions are promoted, and the electrical conductivity of the electrolyte increases. As a result of this, the fill factor and the short-circuit photocurrent are increased. On the other hand, heating adversely affects the charge separation in a semiconductor and the open-circuit photovoltage is expected to decrease. Therefore, with elevation of temperature the cell efficiency curve will pass through a maximum. These regularities were indeed observed in the experiments conducted with cells with CdSe polycrystalline electrodes in a polysulfide ion solution [202]. The cell efficiency was maximum at 50 °C. Also it has been shown that, though the cells get rather heated due to the infrared-light absorption by the solution, all the same the passive cooling of the cell permits of maintaining the admissible temperature conditions provided the cell shape is chosen properly and the walls have good thermal insulation.

6.2.2 Photoanodes made of Ternary I-III-VI₂ Compounds

Of the new materials for photoanodes, studied until recently, the ternary compound $CuInSe_2$ appears to be promising [203]. This is a n-type semiconductor with a forbidden bandwidth of 1 eV. Thanks to direct transitions, it exhibits a good light-absorbing capacity (the light absorption coefficient exceeds 10^5 cm^{-1}). It is used as a photoanode in polysulfide (S^{2-}-S_2^{2-}) and polyiodide (I^--I_3^-) electrolytes, and solar energy can be converted into electrical energy with an efficiency of 3.5 % practically without giving any additional treatment to the crystal surface. $CuInSe_2$ as such is not very stable in the electrolyte solution. For example, dissolution or passivation takes place in I^- solutions and photocurrent decreases rapidly (Fig. 81 b, curve 2).

However, it has been found that the efficiency and stability of this electrode increase radically if its surface gets a special treatment. First, the crystals are etched in methanolic solution of bromine. This affords removal of surface states and of Fermi-level pinning caused by these states. Further the surface can be treated by any one of the following procedures.

A. A layer of indium is electrodeposited on the $CuInSe_2$ surface and the electrode is heated in air at 90 °C until indium changes into indium oxide. At the same time, the near-surface layer of the material is depleted of copper. The formed solid-state structure $CuInSe_2$/indium oxide behaves like a Schottky diode, and the electrode would be thought to have a heterojunction (see Sect. 3.3.3). But

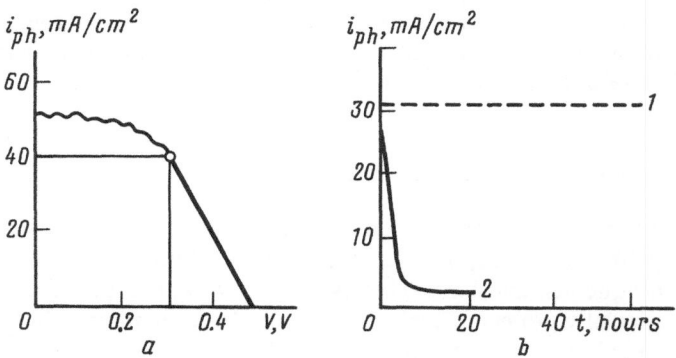

Fig. 81. Working characteristic (a) and dependence of photocurrent on time (b) for a liquid-junction solar cell with a $CuInSe_2$ photoanode, platinum cathode, and polyiodide electrolyte [182]
(a) $2\,M\,KI + 2 \times 10^{-2}\,M\,I_2 + 2 \times 10^{-2}\,M\,CuI + 2 \times 10^{-2}\,M\,In_2(SO_4)_3$ solution, electrode covered with indium oxide layer, 100 mW/cm²; (b) curve 1 - solution, as in (a), curve 2 - $2\,M\,KI + 2 \times 10^{-2}\,M\,I_2$ solution, no indium oxide layer on the electrode (Reprinted by the permission of the publisher, The Electrochemical Society, Inc.)

the oxide layer is porous and the electrolyte penetrates through it to the surface of $CuInSe_2$. Nevertheless, the electrode is chemically quite stable in a polyiodide electrolyte containing additions of Cu^+ and In^{3+} [181] (see Fig. 81 b, curve 1), the reasons being not yet fully known.

B. Photoelectrochemical etching of $CuInSe_2$ in the working solution ($1 M I^- + 5 \times 10^{-2} M I_3^- + 2 M HI + 2 \times 10^{-2} M CuI$) results in the formation of a $CuISe_3$ layer containing elemental selenium, on the surface of $CuInSe_2$. This material is a p-type semiconductor (forbidden bandwidth equals about 2 eV), therefore, an anisotypic (n-p)-heterojunction appears in the photoanode. The outer layer protects the electrode against photocorrosion [204].

Either of the described experimental approaches permits to obtain a regenerative photoelectrochemical cell with the following parameters: efficiency 12 %, open-circuit photovoltage 0.49 V, short-circuit photocurrent 39 mA/cm², fill factor 0.62 for an irradiation power density of 100 mW/cm². Using procedure A, the authors of Ref. [182] obtained the same value of efficiency for polycrystalline samples (crystal size 1–2 mm); the working characteristic of such a cell is presented in Fig. 81 a.

The stability of $CuInSe_2$ in polyiodide (as well as polysulfide) electrolytes, as follows from the calculated values of decomposition potentials [205], is partially thermodynamic and partially kinetic in nature. It is significant that, unlike the CdS anode, selenium is not replaced by sulfur in the near-surface layer when the photoanode operates in the sulfide ions solution.

A preliminary idea of service life of the cell can be obtained from the following data: during 3 months operation the photocurrent decreased only by 5 % [206].

Of the other ternary chalcogenides, $CuInS_2$ has been studied. It in many respects resembles $CuInSe_2$, but is inferior to the latter in energy conversion efficiency. Of great interest are p-type semiconductor compounds suitable for use as photocathodes: $HgIn_2Te_4$ and $CdIn_2Te_4$. In the cells containing solutions of Fe and Cr complex compounds as a redox electrolyte, the light-to-electrical energy conversion efficiency amounts to 2–3 %, and the operation of electrodes is very stable [207]. The studies of such cells are yet in the budding state.

6.2.3 Transition Metals Dichalcogenide Photoelectrodes

The semiconductors with d → d transitions, to which the transition-metal dichalcogenides belong, are, with a few exceptions, as mentioned in Sect. 4.3, prone to photodecomposition in aqueous solutions. That is why they cannot be used as photoelectrodes if not specially protected against corrosion. In regenerative photoelectrochemical cells where such a protection is provided by the redox system (I^-–I_3^-, more rarely Br^-–Br_3^- or Fe^{2+}–Fe^{3+}) present in the solution, the transition-metal dichalcogenides function stably: MoS_2, $MoSe_2$, WS_2, and n-type WSe_2 act as photoanodes, and p-type WSe_2, as photocathode. The forbidden bandwidth (1–1.7 eV) of these materials is best suited for solar energy conversion.

Though comparatively intensive studies of these materials have not yet been made, nevertheless it can be concluded [208] that selenides are superior to sulfides as photoelectrodes and individual compounds outperform the mixed compounds such as WSe_xS_{2-x} or $W_yMo_{1-y}Se_2$. A maximum solar energy conversion efficiency of 12 % (see Table 6.1) has been attained in cells with a WSe_2 photoanode in a polyiodide electrolyte. The service life of certain cells exceeded 1 year.

The main distinguishing feature of transition-metal dichalcogenides - anisotropy of surface, electrochemical and other properties - is attributed to the "layered" nature of the crystal lattice. As already mentioned in Sect. 4.3.2, the surface parallel to the closely packed layers (the so-called Van der Waals surfaces) feature least density of defects and are noted for their close-to-the-"ideal" electrochemical behavior. In contrast, the surfaces inclined to the layers have a large number of defects ("steps") which act as recombination centers. Therefore, the electrodes with "stepped" surfaces have moderate electrochemical characteristics. Besides, the specific adsorption of ions at these steps is generally enhanced [209] and this tends to increase the potential drop in the Helmholtz layer and, hence, to "pin" the Fermi level at the surface, which lowers the photopotential (see Sect. 4.3.2).

Thus, in manufacturing electrodes, one mainly tries to obtain an "ideal" (i.e., Van der Waals) surface. To accomplish this, the harmful effect of the remaining steps is eliminated by subjecting the electrode surface to chemical processing: surface polymerization, e.g., of o-phenylenediamine [210], adsorption of organic substances (for example, ethylenediamine tetraacetic acid) [211], and of the ions of certain metals, etc. Thanks to elevated chemical activity of steps, all these processes are directed just to these steps. Fixing chemical substances on the steps deactivates them. Unfortunately, the action of these processes is but temporary.

On the whole, this class of electrodes invites further investigation into the crystals preparation and their surface treatment techniques.

6.2.4 Electrodes made of Group III–V Compounds

From the end of 1970s when the n-GaAs| $0.8\,M$ K_2Se-$0.1\,M$ K_2Se_2-$1\,M$ KOH|C liquid-junction solar cell was developed by Heller et al. (in the Bell Laboratories, USA) until recently this cell had remained the most efficient photoelectrochemical cell. Gallium arsenide is one of the most suitable materials for solar energy conversion, but it readily corrodes in aqueous solutions. Photocorrosion of the GaAs anode is almost completely eliminated in an about $1\,M$ K_2Se solution: at an irradiation power density of $100\,mW/cm^2$, the decrease in electrode thickness after one year of operation does not exceed several microns [173].

The cause of high cell-efficiency, besides good electrochemical reversibility of the selenide-polyselenide ion system and the specially developed method for chemical etching of GaAs (in the unstirred H_2O_2-H_2SO_4-H_2O solution) which yields a hillocked black but not a light-reflecting surface, was the decrease in surface recombination velocity. To this end, the GaAs surface is treated with solutions of certain metals ions, of which Ru^{3+} (particularly in combination with Pb^{2+}) [173], $Co(NH_3)_5(H_2O)^{2+}$ [212] and Os^{3+} [174] have been found to be most effective. Thus, keeping the electrode in $0.01\,M$ Os^{3+} solution for 1 minute permitted to obtain a photoelectrochemical cell with an efficiency of 15 % - maximum efficiency attained to date in liquid-junction solar cells. At a radiant power density of

$100 \ mW/cm^2$, the cell has the following characteristics: open-circuit photovoltage 0.78 to 0.81 V, short-circuit photocurrent 24-26 mA/cm^2, fill factor 0.65-0.75. On passing 3000 C/cm^2 electricity through the cell, which exceeds 150 times the amount required for complete dissolution of electrode, the photocurrent decreased only by 5 %.

Replacing the GaAs by a solid-solution $GaAs_{1-x}P_x$ electrode allows a slight increase in the photovoltage: up to 1.01 V (at x = 0.25). The efficiency of the cell containing an acetonitrile solution of the ferrocene-ferricenium system equaled 13 %; this is as much as in the cell with aqueous solution of the selenide-polyselenide ion system [213].

All the aforementioned data relate to monocrystalline GaAs samples (including epitaxial films). By using polycrystalline electrodes one succeeds also in attaining high energy conversion efficiency (7.8 %) [173]. With polycrystals the most acute problem is to reduce the recombination velocity which is very intense just at the intercrystalline boundaries. Figure 82 shows how drastically the adsorption of Ru^{3+} (followed by diffusion of Ru along intercrystalline boundaries) improves the working characteristic of the cell with a polycrystalline GaAs electrode. The efficiency of such a cell increases by 4 times. A polycrystalline electrode is very stable too: upon passing 11,000 C/cm^2, the photocurrent decreases only by 15 % [184].

Indium phosphide, the most effective material for photocathodes, also belongs to the Group III-V semiconductor compounds. The photoelectrochemical cell p-InP (monocrystal) | VCl_2-VCl_3-HCl | C has an efficiency of 15 % [173]. Since it operates at sufficiently negative potentials, the indium phosphide cathode does not corrode. To decrease the surface recombination velocity the cathode surface is covered with a thin (0.5-1.0 nm) layer of oxide; this can be achieved also by treating the InP surface with silver salt solutions (e.g., $KAg(CN)_2$). An efficiency of 6.7 % has been attained in a cell with a polycrystalline photoanode (InP films deposited on the graphite substrate, the grain size being 0.5-1 μm) [214].

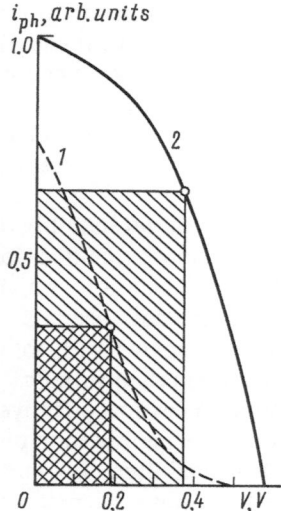

Fig. 82. Working characteristic of n-GaAs | K_2Se-K_2Se_2-KOH | C liquid-junction solar cell with a polycrystalline electrode (GaAs film, average grain size 9 μm, on graphite substrate) [173]
1 - prior to treatment; 2 - after treating with Ru^{3+} and diffusion of Ru along intercrystalline boundaries
Hatched areas are proportional to the output electrical power (Reprinted with permission from Account of Chemical Research. Copyright (1981) American Chemical Society)

Combining a p-type indium phosphide photocathode with a n-type CdSe photoanode described in Sect. 6.2.1 made it possible to make a liquid-junction solar cell with two photoelectrodes: n-CdSe|2 M Na$_2$S-1 M S-1 M NaOH|p-InP. The photopotentials of electrodes are summed up; that is why the cell photovoltage is as high as 1.15 V. Its other characteristics are: short circuit current 24 mA/cm^2 (at 100 mW/cm^2), fill factor 0.4, efficiency 5.5 % [215].

A model has been worked out to explain the action of Ru^{3+} and other adsorbing ions on the surface recombination velocity [173]. According to this model, recombination is caused mainly by the presence of "weak" bonds on the semiconductor surface (or at the intercrystalline boundaries). Saturation of these bonds with strongly adsorbed particles rearranges the system of surface levels: they shift from the middle of the forbidden band, where the levels are most effective in the recombination process, towards its edges and cease to be recombination centers. The oxygen adsorbed on InP plays the same role as Ru^{3+} on GaAs.

6.2.5 Silicon Electrodes

The use of silicon as well as gallium arsenide makes it possible to manufacture high-efficiency solid-state as well as liquid-junction solar cells.

In photoelectrochemical cells, n-type silicon is most often used as a photoanode. Maximum efficiency has been attained with monocrystalline electrodes; electrodes made of amorphous material have also been tested. Reducing agents like iodide or bromide ions in aqueous solution, or ferrocene derivatives in non-aqueous (methanolic) solution are used to protect silicon against photodecomposition.

Because of its high oxidizability, silicon is covered with a several nanometer-thick native SiO$_2$ film. The film adversely affects the photoelectrochemical characteristics of the electrode: at the Si/SiO$_2$ contact, the band bending in silicon is small, therefore the open-circuit photopotential does not exceed 0.4 V. For increasing the photopotential, a special method [216] was developed to replace the native SiO$_2$ film by a controlled-property oxide ensuring both large band bending and fast electron transfer at the interphase boundary. To this end, the sample after chemical etching is immediately placed in a vacuum chamber and the surface is covered with a MgO-Li$_2$O layer about 1 nm in thickness. Thereafter, the sample is annealed at 300–600 °C for 3–6 minutes to eliminate structural defects. In the same manner a 1-nm thick platinum layer is deposited on the oxide layer, which ensures catalytic activity of the electrode. The thin oxide layer does not prevent tunnelling of charges from silicon to solution. The ionic nature of bonds is more pronounced in MgO-Li$_2$O than in SiO$_2$. Hence, a negative charge is built into the oxide, which causes the positive space charge in silicon to grow, i.e., the band bending to increase. That is why the open-circuit photopotential rises from 0.4 to 0.54–0.57 V. In a photoelectrochemical cell containing polybromide or polyiodide electrolyte, the solar energy conversion efficiency reaches 13 % (short-circuit photocurrent 32 mA/cm^2, fill factor 0.69–0.74) [217]. The cell can withstand passing of a charge of up to 20,000 C/cm^2.

The choice of thickness (and, of course, nature) of oxide is of critical importance for photoelectrode characteristics. As the oxide layer thickness is increased (over 2-4 nm), the charge transfer between the semiconductor and the solution is hindered and the photocurrent and the fill factor decrease [218].

In photoelectrochemical cells with a silicon photoanode in $0.2\,M$ methanolic solution of (1-hydroxyethyl)ferrocene (containing also 0.5×10^{-3} mol/l of the oxidized form), the efficiency reached 10 % (open-circuit photovoltage 0.53 V, short-circuit current 20 mA/cm^2) [178].

The forbidden bandwidth of amorphous silicon exceeds that of crystalline silicon: up to 1.8 eV against 1.11 eV. Therefore, the photovoltage in the cells with hydrogenated amorphous silicon (a-Si:H) electrodes is higher: 0.8 V in aqueous solution containing the $Eu^{2+}-Eu^{3+}$ redox system, 0.75-0.85 V in methanolic solution containing the ferrocene-ferricenium system [179, 219]. But their efficiency is small (about 3 %), the reasons being much higher electrical resistivity of the electrode material, increased recombination velocity, etc. Thin-film electrodes are produced by the chemical vapor deposition technique on a stainless steel substrate, with a highly doped a-Si:H sublayer. The latter ensures good ohmic contact.

6.3 Liquid-Junction Cells for Storage of Energy

The electrical energy generated by liquid-junction solar cells can be used by converting it directly into useful work or heat, or stored by converting it into chemical energy with the aid of traditional secondary cells or electrolyzers. The latter case, i.e., "solar cell + electrolyzer" plant is dealt with in detail in Sect. 3.4 by considering solid-state (silicon) solar cells as an example. This approach is applicable, of course, to liquid-junction solar cells, too. Indeed, the optimization method described in Sect. 3.4.2 was used in Ref. [220] in making a unit for producing hydrogen at the cost of solar energy. This unit consists of a liquid-junction solar cell with a thin film $CdSe_{0.65}Te_{0.35}$ photoanode in sulfide-polysulfide electrolyte (efficiency 1.5 %) and an electrolyzer with a solid polymer electrolyte (similar to that described in Sect. 3.4.1). The optimized unit ensures conversion of electrical energy generated by the solar cell into chemical energy with an efficiency of 85 %; the sun-to-hydrogen efficiency equals 1.3 %.

Besides this traditional method, numerous systems have been developed for conversion and storage of solar energy, in which a regenerative photoelectrochemical cell is made integral with a storage device. Such are, for example, the photoelectrochemical storage cells.

If in a regenerative photoelectrochemical cell the anode is separated from the cathode by a diaphragm which allows electric current to pass through it but is impermeable for the redox system components, then the cell changes into a photoelectrochemical storage cell [221]. Its scheme is shown in Fig. 83. It is expedient to take solutions of two different redox systems in the anode and cathode compartments of the cell such that their chemical potential levels $F_{redox,1}$ and $F_{redox,2}$ do not coincide. This is necessary, first, to raise the output voltage and, second, to use

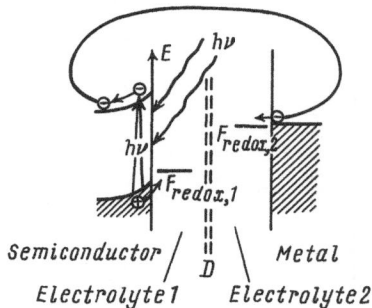

Fig.83. Energy diagram of photoelectrochemical storage cell during charging upon illumination D - diaphragm

more effectively the energy of light-generated carriers (or, as it is sometimes said, to "completely use the semiconductor forbidden bandwidth").

Upon illumination, the photocell gets charged owing to the occurrence of the reactions: $Red_1 + h^+ \rightarrow Ox$ and $Ox_2 + e^- \rightarrow Red_2$. Therefore, the $F_{redox,1}$ and $F_{redox,2}$ levels shift still farther from each other. In the dark with circuit open, the stored energy is conserved for an indefinite period as the diaphragm hinders mixing of these reaction products. When the circuit is made through an external load, the aforementioned reactions proceed in the opposite direction and the cell gives up the stored energy as electrical energy. During discharging, use is generally made of two auxiliary metallic electrodes placed in the anode and cathode compartments of the cell (not shown in Fig. 83). An example of this is the cell n-GaP|$K_3Fe(CN)_6$-$K_4Fe(CN)_6$┊$NiSO_4$|Pt. [222]. When charged in the light, $Fe(CN)_6^{4-}$ gets oxidized to $Fe(CN)_6^{3-}$ and nickel is electrodeposited on platinum; when discharged (in the dark), the processes proceed in the reverse direction.

As the charge capacity per unit volume of the redox system solution is in general small, it was proposed to make use of a flow-type cell so that the total solution volume is not restricted by the interelectrode space. The solutions are stored in special tanks and flow through the illuminated photoelectrochemical cell only during the charging "half-cycle". The solutions of the formed products are sent back to the tanks from where they are returned to the cell for discharge, as may be necessary.

A new version of a photoelectrochemical storage cell is a cell for photoelectro-dialysis (Fig.84) in which water is desalinated at the cost of electrical energy generated by the photoelectrochemical cell, e.g., n-GaAs|Na_2Se-Na_2Se_2-NaOH|Pt. [223, 224]. The cell has five chambers. Chambers 1, 2, and 3 are filled with saline water (in the model experiment, with NaCl solution) and chambers 4 and 5, with working solution of the photoelectrochemical cell ($0.8\,M$ Na_2Se-$0.1\,M$ Na_2Se_2-$1\,M$ NaOH; cf. description of liquid-junction solar cell given in Sect. 6.2.4). The chambers 5 and 1, 2 and 3, and 3 and 4 are separated by cation-exchange membranes, chambers 1 and 2, by an anion-exchange membrane, and chambers 4 and 5, by a "bipolar" electrode pair n-GaAs|Pt. In reality, the chambers 4 and 5 together with the bipolar electrode form a liquid-junction solar cell with short-circuited electrodes. Upon illuminating GaAs, a photovoltage appears in the cell. Under this voltage the ions pass through the membranes, as shown in

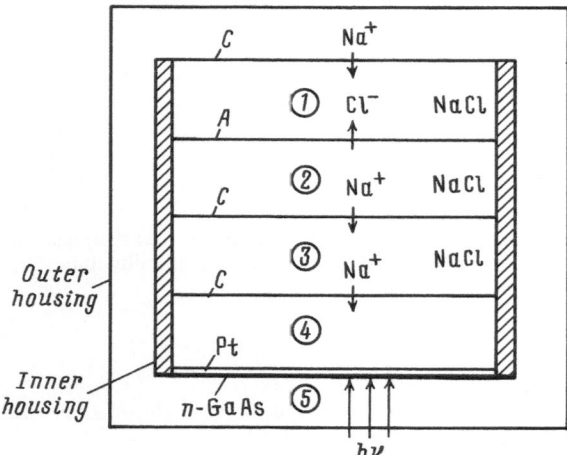

Fig. 84. Schematic of a cell for photoelectrochemical dialysis C – cation-exchange membrane; A – anion-exchange membrane

Fig. 84. As a result, the solution in chamber 2 is depleted both of Na^+ and Cl^- ions, and changes into pure water.

Several connected-in-series semiconductor/metal bipolar systems develop a sufficiently high photovoltage at which electrolysis, e.g., of water can be carried out. In the unit described in Ref. [225], four series connected $CdSe \mid Na_2S-S-KOH \mid CoS$-type liquid-junction solar cells are loaded on the electrolyzer for splitting of water. The "sun-to-hydrogen" efficiency equals about 0.3 %.

All the devices considered in this section (unlike Sect. 6.2) have a rather model-type nature, i.e., demonstrate the principles of using photoelectrochemical cells for the storage of energy, but have not yet been brought up to the development stage that will make possible their practical application. This remains to be a task of the future.

Chapter 7
Protective Coatings for Semiconductor Electrodes

In the regenerative photoelectrochemical cells as well as in the photoelectrolysis cells, described in the foregoing chapters, including those with highest efficiency, the semiconductor electrodes did not have protection as such against photocorrosion in the form of special coatings, surface films, etc. The basis for their stability is merely the fact that the useful reaction in the cell proceeds much faster than the photocorrosion reaction. But special coatings are also given to protect semiconductor electrodes.

7.1 Coatings of Metals and Degenerated Semiconductors

In the past decade attention was repeatedly directed to metal films for protecting semiconductor photoelectrodes against photocorrosion. The protective mechanism of such a film depends upon whether it is continuous or porous. Porous films do not offer mechanical protection; they often accelerate the useful reaction (with the involvement of some component of the solution) competing for minority carriers with the photocorrosion reaction. This mechanism is considered in Sect. 3.2 in the context of operation of semiconductor photocathodes containing metallic islets – catalysts for photoelectrolysis of water.

On the other hand, a continuous metal layer reliably protects the semiconductor by parting it from the solution. Upon illumination, the semiconductor/metal/electrolyte structure, as already mentioned [226], acts as a semiconductor/metal Schottky diode connected in series with the metal/electrolyte boundary. The former provides for photosensitivity of the electrode as a whole, and the latter behaves as if it were an electrical contact to the Schottky diode, which eliminates the problem of lateral electrical resistance in the metal film. In so doing, the film retains its small thickness and therefore does not significantly attenuate light absorption.

In practice, the photopotential, e.g., of n-type GaAs electrode covered with a layer of gold, platinum, or palladium is independent of the reversible potential of the redox system in solution. This photopotential is determined by the band bending at the GaAs/metal boundary, which is dependent only on the difference in the work functions of the contacting phases (i.e., of the semiconductor and metal) but not on the nature of the redox system. The latter, however, determines the kinetics of electron transfer at the metal/solution interface, which, as will be shown in the further discussion, decisively affects the characteristics of the cell as a whole.

On the other hand, this kinetics is almost similar on the metal film covering the semiconductor and on the "macroscopic" electrode of this very metal. Thus, both the interfaces in the electrode covered with the protective film function independently, contributing to the behavior of the photoelectrode (see below).

Under the metal film, GaAs does not practically corrode. But the solar energy conversion efficiency of the cell with such an electrode is roughly half (6 % against 11 %) than that of the cell described in Sect. 6.2.4 with a naked gallium arsenide photoanode. The cause of this decrease in efficiency is that a several nanometer-thick film absorbs just about one-half the incident light energy. Thinner films do not provide anticorrosion protection, because it is hard to make them continuous (poreless).

This compelled the investigators to look for materials other than metals for protective films which, by remaining continuous and good conductors of electric current, would be distinguished by small light absorption. Such materials are oxides whose conductivity type is similar to that of the metal (for example, RuO_2; mixed oxide of ruthenium and titanium (see Sect. 3.5.4); silicides of noble metals) or are semiconductors (or dielectrics) when pure. When heavily doped with donor or acceptor impurities, they attain the conductivity of a degenerated semiconductor (for example, SnO_2-In_2O_3, and others). Thus, a conductive SnO_2-In_2O_3 film several tens of nanometers in thickness transmits up to 90 % light in the visible range [227].

For revealing the contribution of "semiconductor" and "electrolytic" interfaces to photoelectrochemical characteristics of the film-covered electrode, we shall quantitatively describe the behavior of this electrode [36]. Figure 85 shows the energy diagram of an illuminated semiconductor/film/electrolyte solution system.[1] Therein the energy and the electrochemical potential levels are shown for the case when photocurrent is passed through. The Fermi levels of the semiconductor (F) and film (F_l) and the electrochemical potential level (F_{redox}) of the solution containing the redox system Ox-Red do not coincide: the F and F_l levels differ by the value of photopotential in the semiconductor/film Schottky diode; the F_l and F_{redox} levels differ by the value of electrode reaction overvoltage at the film/electrolyte solution interface. The measured electrode photopotential equals:

$$|\varphi_{ph}| = |\varphi_{ph}^d| - \eta \tag{7.1}$$

Here, φ_{ph}^d is the photopotential that appears in the Schottky diode, η is the overvoltage of the electrochemical reaction.[2] In this expression, the photopotentials are

---

[1] The diagram has been drawn for a SnO_2 film which is a n-type degenerated semiconductor (the Fermi level lies close to the conduction band edge or even within the band). But this description is applicable also to the film having purely metallic conduction (in which the Fermi level coincides with Fermi energy).

[2] Strictly speaking, we should have added ohmic losses ($-i_{ph}R$) on the right-hand side of Eq. (7.1). Here, i_{ph} is the photocurrent, R is the total ohmic resistance of the cell. But for the sake of simplicity, R is taken equal to zero in the further discussion.

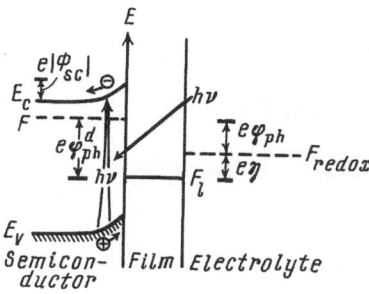

Fig. 85. Energy diagram of n-type semiconductor/oxide film/electrolyte solution system upon illumination [36]

F, F_l, and F_{redox} are electrochemical potential levels, respectively, in semiconductor, film, and solution; $e|\Phi_{sc}|$ – band bending in semiconductor; φ_{ph}^d – photopotential at the semiconductor/film interface; η – overvoltage of electrochemical reaction at film surface; φ_{ph} – measured photopotential of electrode (Reprinted by the permission of the publisher, The Electrochemical Society, Inc.)

given in their absolute values; as in a n-type semiconductor they are negative. In the further discussion, the mathematical sign for "absolute" has been droped for the sake of simplicity.

Let us now turn to the calculation of currents. The photocurrent i_{ph} measured in the cell is the difference of photocurrent i_{ph}^d appearing in the photodiode, and the "leakage current" i_d:

$$i_{ph} = i_{ph}^d - i_d \tag{7.2}$$

The leakage current of diode (i.e., the dark current passing through the semiconductor/film interface in the "forward" direction) equals (see, for example, Ref. [170]):

$$i_d = i_s [\exp(e\varphi_{ph}^d / \gamma kT) - 1] \tag{7.3}$$

where i_s is the saturation current of diode, γ is the quality factor. From Eqs. (7.1) through (7.3) we obtain (at $i_{ph} \gg i_s$):

$$i_{ph} = i_{ph}^d - i_s \exp\left[\frac{e}{\gamma kT}(\varphi_{ph} + \eta)\right] \tag{7.4}$$

The overvoltage η at the "electrolytic" interface is conditionally represented as the sum of two terms. One of them is determined by the diffusion of the components of the redox system in the solution and is expressed (for the simplified case of equal diffusion coefficients for Ox and Red) as:

$$\eta_{dif} = \frac{kT}{ne} \ln\left(\frac{i_{ph} - i_c^{lim}}{i_a^{lim} - i_{ph}}\right) \tag{7.5}$$

Here, n is the number of electrons that participate in the electrode reaction. The limiting diffusion currents of Ox reduction (i_c^{lim}) and Red oxidation (i_a^{lim}) reactions are determined by the conditions of mass transfer in the cell. In Ref. [36], for a particular case of a rotating-disc electrode, these limiting currents have been found by using the theory of this electrode (see, for example, Ref. [147]).

The second term, η_{ct}, is determined by the slowness proper of the electrode reaction and can be found from its kinetic equation (cf. Eq. (1.27)):

$$i_{ph} = i_0[\exp(\alpha ne\,\eta_{ct}/kT) - \exp(-(1-\alpha)ne\,\eta_{ct}/kT)] \qquad (7.6)$$

where α is the transfer coefficient. The exchange current i_0 can be determined, for instance, from the slope of the current-voltage curve near the reversible potential (i.e., at $\eta_{ct} \ll kT/e$):

$$i_0 = \frac{nkT}{e}\,(di_{ph}/d\eta_{ct})_{\eta_{ct}\to 0} \qquad (7.7)$$

Thus, the photopotential measured in the photoelectrochemical cell equals

$$\varphi_{ph} = \frac{kT}{e}\left[\gamma \ln\left(\frac{i_{ph}^d - i_{ph}}{i_s}\right) - \frac{1}{n}\ln\left(\frac{i_{ph} - i_c^{lim}}{i_a^{lim} - i_{ph}}\right) - \frac{1}{\alpha n}\ln\frac{i_{ph}}{i_0}\right] \qquad (7.8)$$

On the right-hand side of Eq. (7.8) the first term is φ_{ph}^d, the second term, η_{dif}, and the third term, η_{ct}. The values of φ_{ph} and i_{ph} can be measured directly in the described photoelectrochemical cell; the characteristics of the Schottky diode (i_s, γ, i_{ph}^d) and of the "electrolytic" contact (i_0, α, n, i_c^{lim}, and i_a^{lim}) are measured on the separately taken corresponding interfaces.

The experimentally found characteristics of a silicon photoanode covered with a 80-nm-thick layer of SnO_2:Sb, in $Fe(CN)_6^{3-}$–$Fe(CN)_6^{4-}$ solution are shown in Fig. 86 for three different illumination intensities. Their inferior quality is very striking: fill factor is as small as 0.28 (at 100 mW/cm^2, curve 3), which results in low (3%) light-to-electrical energy conversion efficiency. Here it is not a matter of properties of the n-Si/SnO_2 junction, but of poor reaction kinetics at the SnO_2/solution interface: the $Fe(CN)_6^{4-}$ oxidation overvoltage for current, conforming to light intensity of 100 mW/cm^2, is as large as 0.3 V (curve 3). In itself the n-Si/

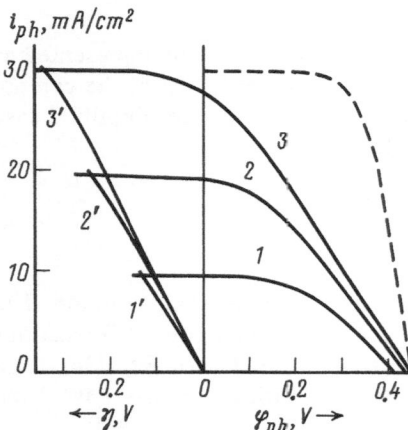

Fig. 86. Measured current-voltage characteristics of n-type Si electrode covered with Sb-doped SnO_2 film in $K_3Fe(CN)_6$-$K_4Fe(CN)_6$-KNO_3 solution [36]
On the right (curves 1, 2, 3) – characteristics of the cell as a whole ($i_{ph} - \varphi_{ph}$); illumination power density: 1 and 1' - 32 mW/cm^2, 2 and 2' - 64 mW/cm^2, 3 and 3' - 100 mW/cm^2; dashed line - dependence of φ_{ph}^d, computed by Eq. (7.1), on photocurrent (for 100 mW/cm^2)
On the left (curves 1', 2', 3') – current-voltage curves of $Fe(CN)_6^{4-}$ oxidation on SnO_2 electrode ($i_{ph} - \eta$) (Reprinted by the permission of the publisher, The Electrochemical Society, Inc.)

SnO_2 Schottky diode has quite satisfactory characteristics (this characteristic computed by Eq. (7.1) is shown in Fig. 86 by a dashed line): fill factor 0.62, efficiency 9 %. The calculation made by Eq. (7.4) using the adjusted Schottky diode characteristics has shown a very close agreement with the experimentally found curve 3, which lends credance to the correctness of the foregoing theory of a film-covered photoelectrode [36].

Thus, though SnO_2 films are rather stable (i.e., can withstand over 1000 C/cm^2) [227], their vulnerable point is small electrocatalytic activity towards the electrode reaction in the photoelectrochemical cell. To improve the electrochemical kinetics, a metal-catalyst (Pt or RuO_2) is added to the film surface (or bulk).

Of other protective oxide coatings, we shall mention RuO_2 on p-Si (spray-deposited in microwave plasma; in the regenerative cell with V^{2+}-V^{3+} system, cf. Sect. 6.2.4, an efficiency of 5.1 % has been attained [228]); tallium oxide on n-Si (electrodeposited film; in the cell with $Fe(CN)_6^{4-}$-$Fe(CN)_6^{3-}$ system the efficiency equals 11 %, open-circuit photovoltage, 0.51 V, short-circuit photocurrent, 33.5 mA/cm^2, fill factor, 0.64 [229]); mixed oxide of ruthenium and titanium on n-Si [117, 230].

Silicon electrodes covered with platinum or iridium silicide films, according to their photoelectrochemical behavior, represent also a Schottky diode loaded on the electrolytic cell with "metallic" electrodes. Indeed, the open-circuit photopotential is independent of the reversible potential of the redox system in solution; on the whole, the electrode characteristics are readily computed by making use of the separately measured characteristics of the silicon/silicide interface in a solid-state system. These films are obtained by evaporating a layer of platinum or iridium on silicon and by heating the sample in air at 250–400 °C. Such a film is catalytically very active in the oxidation reactions, e.g., of halide, Fe^{2+} and $Fe(CN)_6^{4-}$ ions, and also of water. The efficiency runs to 5–8 %, fill factor, up to 0.7. Of the studied coatings, the most stable are silicide films: on passing 10^4 C/cm^2, the photoelectrode characteristics remain almost constant [116, 231].

Finally, the degenerated BP semiconductor is a relatively new material for protective coatings. A 200–500-nm-thick film deposited on n-type silicon or gallium arsenide by the chemical vapor deposition technique permits an efficiency of 2 to 2.75 % in a regenerative cell (containing $Fe(CN)_6^{3-}$-$Fe(CN)_6^{4-}$ solution). The photoelectrode does not appreciably degrade upon passing 30,000 C/cm^2 through the cell [232].

7.2 Coatings of Electrically Conductive Polymers

Electrically conductive polymers – polypyrrole, polyaniline, and polyacetylene – have electronic conductivity as high as 100–1000 ohm^{-1} cm^{-1} for polypyrrole and 10^5 ohm^{-1} cm^{-1} for polyacetylene, which in magnitude approaches the conductivity of metals. At the same time, they are relatively chemically stable in electrolyte solutions under electrode processes. This permits to consider them as candidate-materials of protective coatings for photoelectrodes (see Chap. 14 in Ref. [2] and Ref. [233]).

For a polymer film to well protect the photoelectrode against corrosion, while retaining sufficiently high photoelectrochemical efficiency, it should exhibit good adhesion towards the electrode surface, readily transmit light which is actively absorbed by the semiconductor, and ensure rapid charge transfer from the semiconductor to the reagent in the solution.

The structure of polypyrrole may be represented by the scheme:

where the left-hand part shows the neutral form and the right-hand part, the charged form of polypyrrole. The neutral form (I) is a bad conductor of electric current. Upon oxidation, the conducting form (II) is formed in which one positive charge is generally smeared over 3-4 molecules of pyrrole (such that $x = 1$-2). This positive charge, for the condition that the film as a whole is electrically neutral to hold, is compensated by the charge of anions (X^-) that enter into the film during its oxidation. Such anions may be I_3^-, BF_4^-, ClO_4^-, and others. Oxidation of films and entrapment of anions by the film can be performed in the ambient gas. But a more convenient (and universal) method involves oxidation of film in an electrolyte solution, which follows the process proper of electrolytic growing the polymer coating (see below).

The conduction mechanism of polypyrrole and alike conductive polymers has not yet been fully ascertained. It is quite natural to expect easy transfer of charges along the chains of conjugate double bonds, whereas charges transfer between the ends of these chains by the hopping mechanism. The ions entering into the polymer film perhaps encourage hopping by acting as "bridges" between the high conductivity portions.

Films can be most conveniently grown on the electrode surface by the photoelectrochemical technique. For this purpose, the electrode is placed in monomer solution usually in acetonitrile or other non-aqueous solvent[3], containing additions of an inert electrolyte (which make the solution electrically conductive) and a polymerization initiator, e.g., BF_4^-, necessary potential is established and the electrode is illuminated. Polymerization starts and occurs mainly on the electrode surface but not in the solution bulk. After growing, the film (usually up to 1 μm in thickness) is subjected to electrochemical oxidation.

In the photoelectrochemical behavior of semiconductor electrodes covered with polypyrrole (or the alike) films, one must single out two key points that shed light upon the protective action mechanism:
1. the flat band potential of the semiconductor does not change when a film is deposited on it;
2. in solutions of different redox systems the electrode photopotential is a linear function of the reversible potential of the system.

---

[3] A small (of the order of 0.1 to 1 %) addition of water often promotes deposition of polypyrrole.

Both these conditions do not enable one to consider a polymer film as a continuous conductive coating similar to metallic films described in the previous section. The semiconductor/film junction in electrolyte does not act as Schottky diode. The point is that electrically conductive films are porous and therefore get impregnated with electrolyte solution which penetrates deep into the film, i.e., right up to the semiconductor surface. It is just the semiconductor/electrolyte interface (at the bottom of pores) that determines the flat band potential and the photopotential. Thus, one cannot talk about mechanical isolation of electrode from electrolyte. The protective action of films is most probably based on that polymer, owing to its high electrical conductivity and fast kinetics of the oxidation reaction of solute on the polymer surface, readily transfers the light-generated holes from the semiconductor into solution, i.e., much faster than they are consumed for semiconductor photodecomposition on the portions not covered by the polymer. To this fast transfer contributes also that the true surface area of the polymer is very large because of its porous structure.

Thus, the polymer film protects the semiconductor against corrosion according to the electrocatalytic mechanism, just as this happens on the electrodes having metal-catalyst islets on the surface (cf. Sect. 3.2): namely, by accelerating the useful reaction such that the relative corrosion rate is found to be negligible.

Films of electrically conductive polymers – polypyrrole and also polyaniline [234], polyacetylene [235], and polythiophene [236] – were used to protect Si, GaAs, GaP, CdS, and CdSe electrodes (both photoanodes and photocathodes) in aqueous solutions. They indeed enable the flow of photocurrent to be maintained over a rather long period – sometimes tens of hours – without appreciable decrease in the photocurrent, while the unprotected electrode under similar conditions becomes inoperative (the photocurrent decreases down to zero) in 1–2 minutes (Fig. 87). Nevertheless, the protective ability of polymer films is comparatively poor. The cause of this is peeling of film either due to its weak initial adhesion to the semiconductor or due to the corrosion of substrate under the film, which also decreases adhesion. For more reliable fixation of film to the semiconductor, it was proposed to deposit on the semiconductor surface, prior to electropolymerization, a small amount of metal (e.g., 1–2 nm platinum) [237] or a monolayer of N-(3-trimethoxysilil)propylpyrrole:

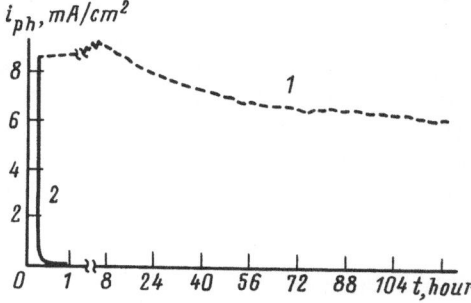

Fig.87. Stabilizing n-type polycrystalline Si electrode by polypyrrole film [245]. Dependence of short-circuit photocurrent on time 1 – electrode covered with film; 2 – naked electrode $FeSO_4$–$FeNH_4(SO_4)_2$–Na_2SO_4 solution (pH 1) (Reprinted with permission from Journal of American Chemical Society. Copyright (1981) American Chemical Society)

which acts like an "anchor" that fastens the polypyrrole film to the semiconductor. Usually films stick more rigidly to polycrystalline rather than to monocrystalline semiconductors, because the former have rough surfaces which favors better adhesion of film to the substrate.

In cells with n-type silicon anodes covered with polypyrrole or with n-type gallium arsenide covered with polythiophene, the light energy conversion efficiency was reported to be as high as 5 % [236, 237]. But in the majority of cases the attained values of efficiency did not exceed 1–2 %. This low efficiency is attributed partly to these polymers, e.g., polypyrrole, films of which rather heavily absorb light in the visible and infrared regions, i.e., the light which is absorbed well in the semiconductors. Besides, the slow kinetics of electrochemical reactions on the surface of polymer films also limits the conversion efficiency and acts as an obstacle to their more common use. In order to catalyze the electrode processes in photoelectrochemical cells, catalytically active admixtures, for example, RuO_2 particles, are implanted on the film surface (sometimes, in the film bulk).

For the same purpose, redox systems like bipyridyl and ferrocene are introduced into the films of certain polymers. These systems, by acting as carriers, accelerate the useful reaction. Thus, it has been shown [238] that in a cell with a n-type GaAs photoanode covered with a polystyrene film containing groups of the Ru(III) bipyridyl complex, and Fe^{2+}–Fe^{3+} solution, the solar energy conversion efficiency is as large as 12 % (photovoltage 1.09 V and fill factor 0.71). But this is essentially the kind of electrode with a chemically modified surface considered in the following section.

7.3 Electrodes with a Chemically Modified Surface

Modification[4] of a semiconductor electrode surface with the aid of a charge carrier (mediator) monolayer attached to the surface has found wide use in the last few years [239]. Generally, a well-reversible redox system acts as a charge carrier. The overvoltage of electrochemical reactions occurring in the monolayer of reagents attached to the surface is less compared to the same reaction proceeding with the involvement of dissolved reagents. As a result, the contribution of the useful process to photocurrent increases and that of photocorrosion, decreases.

Strictly speaking, chemically modified electrodes are not electrodes with a protective coating; they are but conditionally placed in this section. On the other hand, the amount of charge carrier on the electrode surface often exceeds one monolayer, therefore it forms a polymer-type film described above. Such a film also can contain metal-catalyst particles. For this reason, it is difficult to establish a strict line of demarcation between these two classes of electrodes.

As charge carriers, use is generally made of organic systems in which the oxidized (Ox) and reduced (Red) forms differ by one electron – these are first of all

---

[4] Sometimes, the term derivatization is also used in the literature.

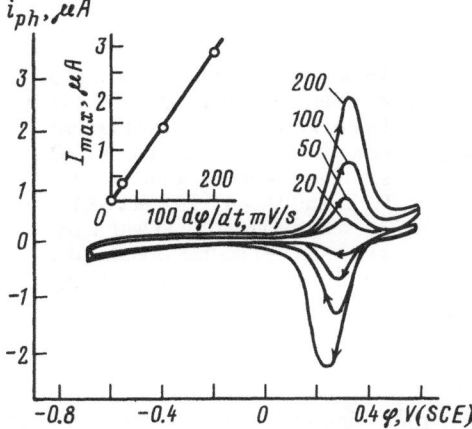

Fig. 88. Cyclic current-voltage curve of illuminated n-type Si electrode (modified with ferrocene) in acetonitrile solution of tetrabutyl ammonium perchlorate [240] Potential scan rate ($d\varphi/dt$, mV/s) is indicated on the curves

Inset figure - plot of peak current (I_{max}) against potential scan rate (Reprinted with permission from Interfacial Photoprocesses. Copyright (1980) American Chemical Society)

ferrocene-ferricenium (and its derivatives) and also other metallocenes, viologens, bipyridyl, etc. Modification of the electrodes consists in attaching a charge carrier to the electrode surface with the aid of chemisorption forces or chemical bonds. In the former case, the semiconductor to be modified is simply immersed into the modifier solution in a suitable solvent and is kept there for some time. In the latter case, a chemical reaction is carried out. In the course of this reaction, a bond, e.g., of type -O- or -NH- is formed between the semiconductor surface atom and the charge carrier molecule. Modification is done by using silane or siloxane compounds and surface hydroxyl groups of the semiconductor:

$$\text{(surface)-OH} + R_1R_2R_3SiCl \rightarrow \text{(surface)-O-}SiR_1R_2R_3 + HCl \tag{7.9a}$$

or

$$\text{(surface)-OH} + R_1R_2R_3Si\text{-}OR_4 \rightarrow \text{(surface)-O-}SiR_1R_2R_3 + R_4OH \tag{7.9b}$$

The amount of the surface-attached substance is usually in the order of 10^{-8} mol/cm². The typical current-voltage curve of an illuminated[5] modified electrode in an indifferent electrolyte solution (Fig. 88) provides an idea of the oxidation and reduction kinetics of the charge carrier in the surface layer: the reaction proceeds at a fast (the peak current is proportional to the potential scan rate) but yet at a finite rate (the potentials of the anodic and cathodic current peaks slightly move apart with increasing scan rate). Such a current-voltage curve can be obtained repeatedly: the charge carrier is firmly attached to the electrode surface.

If the solution contains the same redox system (e.g., ferrocene/ferricenium) which is attached to the surface, then charge exchange takes place between the solution and the surface layer as follows: ferricenium attached to the surface oxidizes the dissolved ferrocene and gets reduced itself, and so on. An important point is that the substance in the surface layer exchanges charges with the electrode more easily than the substance from the solution, and therefore acts as catalyst in

---

[5] Practically there is no dark current.

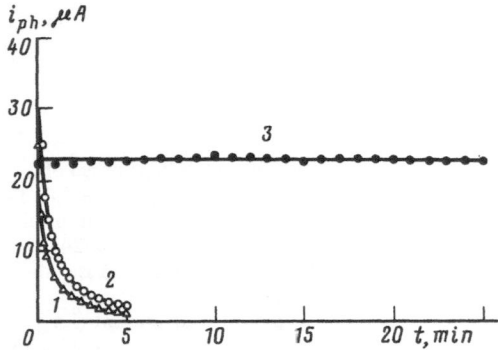

Fig. 89. Stabilization of n-type Si photoanode by modifying its surface [240] Plot of anodic photocurrent against time: 1 - naked Si in $0.1\,M$ NaClO$_4$ solution; 2 - naked Si in $0.1\,M$ NaClO$_4$ + $4 \times 10^{-3}\,M$ Fe(CN)$_6^{4-}$; 3 - solution as in (2), Si with ferrocene-modified surface (Reprinted with permission from Interfacial Photoprocesses. Copyright (1980) American Chemical Society)

the electrode reaction. Improving the electrode kinetics tends to increase the energy conversion efficiency in photoelectrochemical cells. Thus, in a cell with a usual n-type GaAs photoanode and ferrocene-ferricenium solution, the energy conversion efficiency equals 3 %. But on modifying the surface of this very electrode with ferrocene, the cell efficiency increases up to 5 % [240].

The charge exchange between the surface layer of the carrier and the solution takes place as well if the redox system in the solution is different from that on the surface, for example, if silicon modified with ferrocene is immersed in ferrocyanide solution (Fig. 89). The latter as such poorly protects the silicon photoanode against decomposition because of an insufficiently large oxidation rate of Fe(CN)$_6^{4-}$. Therefore, photocurrent in the Fe(CN)$_6^{4-}$ solution decreases (due to the growth of the insulating SiO$_2$-film on the Si surface) almost as rapidly as in the solution not containing ferrocyanide (cf. curves 1 and 2). But if the silicon surface is modified with ferrocene, then the light-generated holes are transferred at a high rate from the electrode to Fe(CN)$_6^{4-}$ ions. First, the surface-attached ferrocene gets oxidized and the formed ferricenium oxidizes Fe(CN)$_6^{4-}$ to Fe(CN)$_6^{3-}$, regenerating surface ferrocene, and so on. Therefore, oxidation of Si to SiO$_2$ actually does not take place, and the photocurrent remains constant (Fig. 89, curve 3).

The given example is fascinating in that it shows how ferrocene, insoluble in water, can be used to catalyze the electrode reaction occurring in aqueous solution.

Catalytically active metals (Pt, Pd, Rh) are implanted in the surface layer of the charge carrier to further increase the electrode reaction rate.

Chemically modified electrodes are used in regenerative photoelectrochemical cells (for example, with iodide electrolyte) as well as in photoelectrolysis cells (e.g., for obtaining hydrogen from water and for electroreduction of CO$_2$) [241]. The efficiency of such cells, however, does not exceed 1-2 % as a rule. Perhaps, this is due to not complete elimination of kinetic limitations and also due to insufficiently large band bending in the modified semiconductor. Besides, not very good reliability of the carriers attached to the surface is an obstacle to extensive application of chemically modified electrodes in photoelectrochemical cells.

Conclusion

By the state of development, the photoelectrochemical method of solar energy conversion occupies an intermediate place between the thermochemical and photochemical methods. In fact, while thermochemical conversion is now realized on a semi-industrial scale in thousand-kilowatt solar power stations and photochemical conversion as a whole has not yet exceeded the stage of making models of photochemical processes, the photoelectrochemical studies have lead to the creation of efficient cells which, if the situation of energy resources becomes more strained, could even now find practical application. Among them are liquid-junction solar cells with polycrystalline electrodes; certain types of photoelectrolysis cells, in particular, for splitting halide acids with the aid of film-protected electrodes; and "solar cell + electrolyzer" plants. These cells, however, yet demand solutions to a number of engineering and manufacturing problems related to sealing of cells, selection of structural materials, and others.

Other trends are being intensively developed and shall undoubtedly result in the creation of competitive solar energy converters. Here, mention must be made first of all of the study of microheterogeneous (suspension and colloidal) systems, photointercalation processes, and also of dimensional quantum effects, for example, in semiconductor microparticles, in the so-called electrodes with super lattices [242], and others.

The entire material of this book points, to our mind, to the importance of the enormous possibilities of the photoelectrochemical conversion of solar energy.

References

1. Pleskov YuV, Gurevich YuYa (1986) Semiconductor Photoelectrochemistry. Consultants Bureau, New York, London
2. Grätzel M (ed) (1983) Energy Resources through Photochemistry and Catalysis. Academic Press, New York, London
3. Finklea HO (ed) (1988) Semiconductor Electrodes. Elsevier, Amsterdam, Oxford, New York, Tokyo
4. Kalyanasundaram K (1985) Solar Cells 15:93
5. Bockris JO'M (1980) Energy Options. Australia and New Zealand Book Company, Sydney, Melbourne, Brisbane, Auckland
6. Bockris JO'M (1975) Energy. The Solar-Hydrogen Alternative. Australia and New Zealand Book Company, Sydney, Melbourne, Brisbane, Auckland
7. Semenov NN, Shilov AE (eds) (1985) Preobrazovanie solnechnoi energii. Nauka, Moscow
8. Newman JS (1973) Electrochemical Systems. Prentice Hall, Englewood Cliffs, NJ
9. Gurevich YuYa, Pleskov YuV, Rotenberg ZA (1980) Photoelectrochemistry. Consultants Bureau, New York, London
10. Gerischer H, Spitler MT, Willig F (1980) in: Bruckenstein S, Miller B, McIntyre JDE, Yeager E (eds) Electrode Processes. The Electrochemical Society, Princeton, NJ, p 115
11. Albery WJ (1982) Acc. Chem. Res. 15:142
12. Smith RA (1978) Semiconductors. Cambridge University Press, Cambridge
13. Gerischer H, Eckardt W (1983) Appl. Phys. Lett. 43:393
14. Parsons R (1985) in: Bard AJ, Parsons R, Jordan J (eds) Standard Potentials in Aqueous Solutions. Marcel Dekker, New York, Basel, p 13
15. Trasatti S (1982) J. Electroanalyt. Chem. 139:1
16. Gurevich YuYa, Pleskov YuV (1982) Elektrokhimiya 18:1477
17. Trasatti S (1986) Pure and Appl. Chem. 58:955
18. Pleskov YuV (1987) J. Phys. Chem. 91:1691
19. Reiss H, Heller A (1985) ibid. 89:4207
20. Frumkin AN (1982) Potentsialy nulevogo zaryada. Nauka, Moscow
21. Yoneyama H, Hoflund GB (1986) Progr. Surface Sci. 21:5
22. Gerischer H (1984) J. Electrochem. Soc. 131:2452
23. Sculfort JL, Triboulet R, Lemasson P (1984) ibid. 131:209
24. Gerischer H (1979) in: Seraphin BO (ed) Solar Energy Conversion. Springer, Berlin, Heidelberg, p 115
25. Pleskov YuV (1967) Electrokhimiya 3:112
26. Van Meirhaeghe RL, Cardon F, Gomes WP (1985) J. Electroanal. Chem. 188:287
27. Lorenz W, Herrnberger H (1986) ibid. 197:167
28. Jaeger CD, Gerischer H, Kautek W (1982) Ber. Bunsenges. Phys. Chem. 86:20
29. Brudel M, Janietz P, Landsberg R (1987) J. Electroanal. Chem. 86:20
30. Nagasubramanian G, Wheeler BL, Bard AJ (1983) J. Electrochem. Soc. 130:1680
31. Popkirov G, Schindler RN (1986) Solar Energy Mater. 13:161
32. Damaskin BB, Petrii OA (1983) Vvedenie v elektrokhimicheskuyu kinetiku. Vysshaya shkola, Moscow
33. Albery WJ, Bartlett PN, Hamnett A, Dare-Edwards MP (1981) J. Electrochem. Soc. 128:1492
34. Albery WJ, Bartlett PN (1983) ibid. 130:1699

35. Kautek W, Gerischer H (1982) Electrochim. Acta 27:355
36. Decker F, Fracastoro-Decker M, Badawy W, Doblhofer K, Gerischer H (1983) J. Electrochem. Soc. 130:2173
37. Lorenz W (1985) J. Electroanal. Chem. 191:31
38. Li J, Peter LM (1985) ibid. 193:27
39. Orazem ME, Newman J (1984) J. Electrochem. Soc. 131:2569
40. Cooper G, Turner JA, Parkinson BA, Nozik AJ (1983) J. Appl. Phys. 54:6463
41. Memming R (1987) Ber. Bunsenges. phys. Chem. 91:353
42. Allongue P, Cachet H (1985) J. Electrochem. Soc. 132:45
43. Lincot D, Vedel J (1987) J. Electroanal. Chem. 220:179
44. Vanmaekelbergh D, Gomes WP, Cardon F (1985) Ber. Bunsenges. Phys. Chem. 89:994
45. Gerischer H (1981) in: Cardon F, Gomes WP, Dekeyser W (eds) Photovoltaic and Photoelectrochemical Solar Energy Conversion. Plenum Press, New York, London, p 199
46. Memming R (1988) Topics in Curr. Chem. 143:81
47. Nozik AJ (1981) in: Cardon F, Gomes WP, Dekeyser W (eds) Photovoltaic and Photoelectrochemical Solar Energy Conversion. Plenum Press, New York, London, p 263
48. Fujishima A, Honda K (1971) Bull. Chem. Soc. Japan 44:1148
49. Kunitzky YuA (1985) Elektrodnye materialy dlya pryamykh preobrazovatelei energii. Vishcha shkola, Kiev, Sect. 4:1
50. Bin-Daar G, Dare-Edwards MP, Goodenough JB, Hamnett A (1983) J. Chem. Soc. Faraday Trans. Ser. 1, 79:1199
51. Bolton JR, Strickler SJ, Connolly JS (1985) Nature 316:495
52. Weber MF, Dignam MJ (1984) J. Electrochem. Soc. 131:1258
53. Kühne HM, Tributsch H (1986) J. Electroanal. Chem. 201:263
54. Ang PGP, Sammels AF (1984) J. Electrochem. Soc. 131:1462
55. Kainthla RC, Bockris JO'M (1987) Int. J. Hydrogen Energy 12:23
56. Heller A, Vadimsky RG (1981) Phys. Rev. Lett. 46:1153
57. Aharon-Shalom E, Heller A (1982) J. Electrochem. Soc. 129:2865
58. Heller A, Aharon-Shalom E, Bonner WA, Miller B (1982) J. Amer. Chem. Soc. 104:6942
59. Contractor AQ, Szklarczyk M, Bockris JO'M (1983) J. Electroanal. Chem. 157:175
60. Bockris JO'M, Szklarczyk M, Contractor AQ, Khan SUM (1984) Int. J. Hydrogen Energy 9:741
61. Szklarczyk M, Bockris JO'M (1984) J. Phys. Chem. 88:1808
62. Dominey RN, Lewis NS, Bruce JA, Bookbinder DC, Wrighton MS (1982) J. Amer. Chem. Soc. 104:467
63. Heller A (1985) J. Phys. Chem. 89:2962
64. Aspnes DE, Heller A (1983) ibid. 87:4919
65. Tsubomura H, Nakato Y (1987) New J. Chem. 11:167
66. Popkirov GS, Sakharova AYa, Pleskov YuV (1988) Int. J. Hydrogen Energy 13:681
67. Turner JE, Hendewerk M, Parmeter J, Neiman D, Somorjai GA (1984) J. Electrochem. Soc. 131:1777
68. Kainthla RC, Zelenay B, Bockris JO'M (1987) ibid. 134:841
69. Bard AJ (1982) J. Phys. Chem. 86:172
70. Sammels AF, Ang PGP (1985) US Pat. 4545872
71. Loferski JJ (1981) in: Cardon F, Gomes WP, Dekeyser W (eds) Photovoltaic and Photoelectrochemical Solar Energy Conversion. Plenum Press, New York, London, p 157
72. Morisaki H, Watanabe T, Iwase M, Yazawa K (1976) Appl. Phys. Lett. 29:338
73. Wagner S, Shay JL (1977) ibid. 31:446
74. Osaka T, Hirota N, Hayashi T, Eskildsen SS (1985) Electrochim. Acta 30:1209
75. Kulak AI (1986) Elektrokhimiya poluprovodnikovykh geterostruktur. Universitetskoye izdatelstvo, Minsk
76. Li G, Wagner S (1987) J. Electroanal. Chem. 227:213
77. Ueda K, Nakato Y, Sakamoto H, Sakai Y, Matsumura M, Tsubomura H (1987) Chem. Lett. No. 4:747
78. Bockris JO'M, Dangapani B, Cocke D, Ghoroghchian J (1985) Int. J. Hydrogen Energy 10:179
79. Appleby AJ, Delahoy AE, Gau SC, Murphy OJ, Kapur M, Bockris JO'M (1985) Energy 10:871

80. Esteve D, Ganibal C, Steinmetz D, Vialaron A (1982) Int. J. Hydrogen Energy 7:711
81. Carpetis C, Schnurnberger W, Seeger W, Steeb H (1982) in: Hydrogen Energy Progr. IV, Vol 4. Pergamon Press, Oxford p 1495
82. Koukouvinos A, Ligerou V, Koumoutsos N (1982) Int. J. Hydrogen Energy 7:645
83. Pleskov YuV, Zhuravleva VN, Pshenichnikov AG, Vartanyan AV, Arutynyan VM, Sarkisyan AG, Melikyan VM (1985) Geliotekhnika No. 4:61
84. Salamov OM, Bakirov MYa, Rzaev PF (1986) ibid. No. 4:43
85. Bychkovskii SK, Konev VG, Negreev BM, Strebkov DS, Samoilova LA, Svinarev SV, Trushevsky SN (1986) ibid. No. 4:29
86. Hankock OG (1986) Int. J. Hydrogen Energy 11:153
87. Morimoto Y, Hayashi T, Mayeda Y (1986) in: Hydrogen Energy Progr. VI, Vol 1. Pergamon Press, New York, p 326
88. Steeb H, Weiss HR, Khoshaim BH (1986) ibid. p 406
89. Carpetis C (1982) Int. J. Hydrogen Energy 7:287
90. Carpetis C (1984) ibid. 9:969
91. Kharkats YuI, German ED, Kazarinov VE, Pshenichnikov AG, Pleskov YuV (1986) ibid. 11:617
92. Hammashe A, Bilgen E (1986) in: Hydrogen Energy Progr. VI, Vol 1. Pergamon Press, New York, p 287
93. Chowhudry J (1983) Chem. Eng. (USA) 90:30
94. Nutall LJ, Russel JH (1980) Int. J. Hydrogen Energy 5:75
95. Scaife DE (1980) Solar Energy 25:41
96. Hartig KJ, Getoff N, Rumpelmayer G, Popkirov G, Kotchev K, Kanev St (1986) in: Hydrogen Energy Progr. VI, Vol 2. Pergamon Press, New York, p 546
97. Ioffe MS, Salitra GS, Pivovarov AP, Borodko YuG (1984) Khim. Fizika 3:1012
98. Popkirov GS, Pleskov YuV (1980) Elektrokhimiya 16:238
99. Weber MF, Schumacher LC, Dignam MJ (1982) J. Electrochem. Soc. 129:2022
100. Arutyunyan VM (1985) in: Fotokataliticheskoye preobrazovanie solnechnoi energii, Pt 1. Nauka, Novosibirsk, p 74
101. Schumacher LC, Mamiche-Afara S, Dignam MJ (1986) J. Electrochem. Soc. 133:716
102. Memming R (1984) Progr. in Surface Sci. 17:7
103. Schumacher R, Wilson RH, Harris LA (1980) J. Electrochem. Soc. 127:96
104. Armstrong NR, Shepard VR (1982) J. Electroanal. Chem. 131:113
105. Rajeshwar K (1985) Appl. Electrochem. 15:1
106. Honda K, Fujishima A, Watanabe T (1978) in: Takamura T, Kozawa A (eds) Surface Electrochemistry. Japan Scientific Society Press
107. Nakao M, Itoh K, Watanabe T, Honda K (1985) Ber. Bunsenges. phys. Chem. 89:134
108. Kraitsberg AM, Pleskov YuV, Mardashov YuS (1983) Elektrokhimiya 19:435
109. Kraitsberg AM, Pleskov YuV, Mardashov YuS (1986) in: Electrodynamics and Quantum Phenomena at Interfaces. Metsniereba, Tbilisi, p 518
110. Blasse G, Dirksen GJ, da Korte PHM (1981) Mater. Res. Bull. 16:991
111. Jarrett HS, Sleight AW, Kung HH, Gilson JL (1980) App. Phys. 51:3916
112. Guruswamy V, Murphy OJ, Young V, Hildreth G, Bockris JO'M (1981) Solar Energy Mater. 6:59
113. Pleskov YuV, Krotova MD, Revina AA (1985) Radiat. Phys. and Chem. 26:17
114. Pleskov YuV, Krotova MD, Revina AA (1982) USSR Pat. No. 807672
115. Levy-Clement C, Heller A, Bonner WA, Parkinson BA (1982) J. Electrochem. Soc. 129:1701
116. Fan FRF, Keil RG, Bard AJ (1983) J. Amer. Chem. Soc. 105:220
117. Kraitsberg AM, Pleskov YuV (1987) Elektrokhimiya 33:1113
118. Nakato Y, Iwakabe Y, Hiramoto M, Tsubomura H (1986) J. Electrochem. Soc. 133:900
119. Mackor A (1983) in: Hall DO, Palz W, Pirrwitz D (eds) Photochemical, Photoelectrochemical and Photobiological Processes, Vol 2. D. Reidel Publishing Company, Dordrecht Boston Lancaster, p 8
120. Luttmer JD, Trachtenberg I (1985) J. Electrochem. Soc. 132:1312
121. Luttmer JD, Konrad D, Trachtenberg I (1985) ibid. 132:1054
122. Luttmer JD, Trachtenberg I (1985) ibid. 132:1820
123. White JR, Fan FRF, Bard AJ (1985) ibid. 132:544

124. Calabrese GS, Sobieralski TJ, Wrighton MS (1983) Inorg. Chem. 22:1634
125. Dickson CR, Nozik AJ (1978) J. Amer. Chem. Soc. 100:8007
126. Koizumi M, Yoneyama H, Tamura H (1981) Bull. Chem. Soc. Japan 54:1682
127. Halmann M (1984) J. Electroanal. Chem. 181:307
128. Grinberg VA, Dzhavrishvili TV, Vasiliev YuB, Rotenberg ZA, Kazarinov VE, Maiorova NA (1984) Elektrokhimiya 20:121
129. Halmann M (1978) Nature 275:115
130. Canfield D, Frese KW (1983) J. Electrochem. Soc. 130:1772
131. Taniguchi I, Aurian-Blajeni B, Bockris JO'M (1984) Electrochim. Acta 29:923
132. Tributsch H (1986) in: Modern Aspects of Electrochemistry, Vol 17. Plenum Press, New York, London, p 303
133. Tributsch H (1982) Structure and Bonding 49:127
134. McKinnon WR, Haering RR (1983) Modern Aspects of Electrochemistry, Vol 15. Plenum Press, New York, London, p 235
135. Tributsch H (1980) Appl. Phys. 23:61
136. Tributsch H (1983) Solid State Ionics 9–10:41
137. Kautek W, Gerischer H (1982) Surface Sci. 119:46
138. Kabanov BN, Kiseleva IG, Astakhov II (1981) in: Kinetika slozhnykh elektrokhimicheskikh reaktsii. Nauka, Moscow, p 200
139. Grätzel M (1983) in: Modern Aspects of Electrochemistry, Vol 15. Plenum Press, New York, London, p 83
140. Nozik AJ (1976) Appl. Phys. Lett. 29:150
141. Gerischer H (1984) J. Phys. Chem. 88:6096
142. Hodes G, Grätzel M (1984) New J. Chem. 8:509
143. Ueno A, Kakuta N, Park KH, Finlyason MF, Bard AJ, Campion A, Fox MA, Webber SE, White JM (1985) J. Phys. Chem. 89:3828
144. Enea O, Bard AJ (1986) ibid. 90:301
145. Dunn W, Aikawa Y, Bard AJ (1981) J. Electrochem. Soc. 128:222
146. Albery WJ, Bartlett PN, Porter JD (1984) ibid. 131:2892
147. Pleskov YuV, Filinovskii VYu (1976) The Rotating Disc Electrode. Consultants Bureau, New York, London
148. Rossetti R, Nakahara S, Brus LE (1983) J. Chem. Phys. 79:1086
149. Henglein A (1988) Topics in Curr. Chem. 143:115
150. Nedeljkovich JM, Nanadovich MT, Micic OI, Nozik AJ (1986) J. Phys. Chem. 90:12
151. Gerischer H, Lübke M (1986) J. Electroanal. Chem. 204:225
152. Krasnovsky AA, Brin GP (1962) Doklady Akad. Nauk SSSR 147:656
153. Bulatov AV, Khidekel ML (1976) Izvestia Akad. Nauk SSSR, Ser. Khim. 1902
154. Duonghong D, Borgarello E, Grätzel M (1981) J. Amer. Chem. Soc. 103:4685
155. Gu B, Kiwi J, Grätzel M (1985) New J. Chem. 9:539
156. Borgarello E, Kalyanasundaram K, Okuno Y, Grätzel M (1981) Helv. Chim. Acta 64:1937
157. Gissler W, McEvoy AJ, Grätzel M (1982) J. Electrochem. Soc. 129:1733
158. Meissner D, Memming R, Kastening B, Bahnemann D (1986) Chem. Phys. Lett. 127:419
159. Meissner D, Memming R, Kastening B (1983) ibid. 96:34
160. Gerischer H (1980) Faraday Disc. Chem. Soc. No. 70:416
161. Hauffe K, Stechenmesser R (1985) Chem. Zeitung 109:215
162. Sobczynski A, Bard AJ, Campion A, Fox MA, Mallouk T, Webber SE, White JM (1987) J. Phys. Chem. 91:3316
163. Kalyanasundaram K, Borgarello E, Grätzel M (1981) Helv. Chim. Acta 64:362
164. Thewissen DHMW, Tinnemans AHA, Eluwhorst-Reinten M, Timmer K, Mackor A (1983) New J. Chem. 7:191
165. Fedoseev VI, Savinov EM, Parmon VN (1987) Kinetika i kataliz 28:1111
166. Fox MA (1987) Topics in Curr. Chem. 142:72
167. Hidaka H, Kubota H, Grätzel M, Serpone N, Pelizzetti E (1985) New J. Chem. 9:67
168. Dabestani R, Wang X, Bard AJ, Campion A, Fox MA, Webber SE, White JM (1986) J. Phys. Chem. 90:2729
169. Tricot YM, Emeren A, Fendler JH (1985) ibid. 89:4721
170. Sze SM (1981) Physics of Semiconductor Devices. Wiley, New York

171. Orazem ME, Newman J (1986) Modern Aspects of Electrochemistry, Vol 18. Plenum Press, New York, London, p 61
172. Gale RJ, Dubow J (1981) Solar Energy Mater. 4:135
173. Heller A (1981) Acc. Chem. Res. 14:154
174. Tufts BJ, Abrahams IL, Santangelo PG (1987) Nature 326:861
175. Gronet CM, Lewis NS (1982–1983) ibid. 300:733
176. Miller B, Heller A, Robbins M, Menezes S, Chang KC, Thomson J (1977) J. Electrochem. Soc. 124:1019
177. Heller A, Leamy HJ, Miller B, Johnston WD (1983) J. Phys. Chem. 87:3239
178. Gronet CM, Lewis NS, Cogan G, Gibbons J (1983) Proc. Nat. Acad. Sci. USA, Phys. Sci. 80:1152
179. Gronet CM, Lewis NS, Cogan GW, Gibbons JF, Moddel GR, Weismann H (1984) J. Electrochem. Soc. 131:2873
180. Kline G, Kam K, Canfield D, Parkinson BA (1981) Solar Energy Mater. 4:301
181. Cahen D, Chen YW, Noufi R, Ahrenkiel R, Matson R, Tomkiewicz M, Shen WM (1986) Solar Cells 16:529
182. Razzini G, Peraldo Bicelli L, Scrosati B, Zanotti L (1986) J. Electrochem. Soc. 133:351
183. Weaver NL, Singh R, Rajeshwar K, Singh P, Dubow J (1981) Solar Cells 3:221
184. Hodes G, Fonash SJ, Heller A, Miller B (1984) in: Adv. Electrochem. and Electrochem. Engng. Vol 13. Wiley, New York Chichester Brisbane Toronto Singapore, p 113
185. Fulop GF, Taylor RM (1985) Annu. Rev. Mater. Sci. Vol 15. Palo Alto, Calif., p 197
186. Tenne R, Muller N, Mirovsky Y, Lando D (1983) J. Electrochem. Soc. 130:852
187. Flaisher H, Tenne R, Hodes G (1984) J. Phys. D: Appl. Phys. 17:1055
188. Kovach SK, Vasko AT, Gorodyskii AV, Chornokozha TS (1986) Elektrokhimiya 22:808
189. Inoue T, Watanabe T, Fujishima A, Honda K (1977) J. Electrochem. Soc. 124:719
190. Hodes G, Manassen J, Cahen D (1980) ibid. 127:544
191. Licht S, Tenne R, Flaisher H, Manassen J (1986) ibid. 133:52
192. Licht S (1986) J. Phys. Chem. 90:1096
193. Ellis AB, Kaiser SW, Bolts JM, Wrighton MS (1977) J. Amer. Chem. Soc. 99:2839
194. Noufi R, Kohl PA, Rogers JW, White JM, Bard AJ (1979) J. Electrochem. Soc. 126:949
195. Heller A, Schwartz GP, Vadimsky RG, Menezes S, Miller B (1978) ibid. 125:1156
196. Boudreau RA, Rauh RD (1983) ibid. 130:513
197. Liu CJ, Wang JH (1982) ibid. 129:719
198. Szabo JP, Cocivera M (1986) ibid. 133:1247
199. Minoura H, Negoro T, Kitakata M, Ueno Y (1985) Solar Energy Mater. 12:335
200. Russak MA, Creter C (1984) J. Electrochem. Soc. 131:556
201. Kolbasov GYa, Karpov II, Pavelets AM, Khanat LN (1986) Geliotekhnika No. 2:3
202. Müller N, Cahen D (1983) Solar Cells 9:229
203. Shen WM, Tomkiewicz M, Cahen D (1986) J. Electrochem. Soc. 133:112
204. Menezes S (1986) Solar Cells 16:255
205. Cahen D, Mirovsky Y (1985) J. Phys. Chem. 89:2818
206. Bachman KJ, Menezes S, Kötz R, Fearheily M, Lewevenz HJ (1984) Surface Sci. 138:475
207. Becker RS, Zhow GD, Elton J (1986) J. Phys. Chem. 90:5866
208. Kline G, Kam KK, Ziegler R, Parkinson BA (1982) Solar Energy Mater. 6:337
209. Lewevenz HJ, Gerischer H, Lübke M (1984) J. Electrochem. Soc. 131:100
210. White HS, Abruna HD, Bard AJ (1982) ibid. 129:265
211. Razzini G, Peraldo Bicelli L, Pini G, Scrosati B (1981) ibid. 128:2134
212. Abrahams IL, Tufts BJ, Lewis NS (1987) J. Amer. Chem. Soc. 109:3472
213. Gronet CM, Lewis NS (1984) J. Phys. Chem. 88:1310
214. Heller A (1982) J. Vac. Sci. Technol. 21:559
215. Ang PGP, Sammels AF (1983) J. Electrochem. Soc. 130:1784
216. Howe AT (1983) J. Chem. Soc. Chem. Communs. 1407
217. Howe AT, Fleisch TH (1987) J. Electrochem. Soc. 134:72
218. Canfield DG, Morrison SR (1982) Appl. Surface Sci. 10:493
219. Harrison DY, Calabrese GS, Ricco AJ, Dresner J, Wrighton MS (1983) J. Amer. Chem. Soc. 105:4212

220. Gorodyskii AV, Zhuravleva VN, Karpov II, Kolbasov GYa, Pleskov YuV, Kharkats YuI (1987) Elektrokhimiya 23:1443
221. Ang PGP, Sammels AF (1980) Faraday Disc. Chem. Soc. No. 70:207
222. Yonezawa Y, Okai M, Ishino M, Hada H (1983) Bull. Chem. Soc. Japan 56:2873
223. Murphy GW (1978) Solar Energy 21:403
224. Murphy GW (1983) US Pat. No. 4404081
225. Smotkin ES, Cervera-Marsch S, Bard AJ, Campion A, Fox MA, Mallouk T, Webber SE, White JM (1987) J. Phys. Chem. 91:6
226. Menezes S, Heller A, Miller B (1980) J. Electrochem. Soc. 127:1268
227. Decker F, Melsheimer J, Gerischer H (1982) Israel J. Chem. 22:195
228. Gissler W, McEvoy AJ (1984) Solar Energy Mater. 10:309
229. Switzer JA (1986) J. Electrochem. Soc. 133:723
230. Kraitsberg AM, Pleskov YuV (1989) Elektrokhimiya 25:836
231. Fan FRF, Hope GA, Bard AJ (1982) J. Electrochem. Soc. 129:1647
232. Ginley DS, Baughman RJ, Butler MA (1983) J. Electrochem. Soc. 130:1999
233. Noufi RN (1983) Appl. Phys. Communs. 3:33
234. Noufi R, Nozik AJ, White J, Warren LF (1982) J. Electrochem. Soc. 129:2261
235. Simon RA, Wrighton MS (1984) Appl. Phys. Lett. 44:930
236. Horowitz G, Tourillon G, Garnier F (1984) J. Electrochem. Soc. 131:151
237. Skotheim T, Pettersson LG, Inganäs O, Lundstrom I (1982) ibid. 129:1737
238. Rajeshwar K, Kaneko M, Yamada A (1983) ibid. 130:38
239. Bard AJ (1983) J. Chem. Educ. 60:302
240. Wrighton MS, Bocarsly AB, Bolts JM, Bradley MG, Fischer AB, Lewis NS, Palazzotto MG, Walton EG (1980) in: Wrighton MS (ed) Interfacial Photoprocesses: Energy Conversion and Storage. American Chemical Society, Washington, D.C., p 269
241. Daube KA, Harrison DJ, Mallouk TE, Ricco AJ, Chao S, Wrighton MS (1985) J. Photochem. 29:71
242. Nozik AJ, Thacker BR, Turner JA, Olson JM (1985) J. Amer. Chem. Soc. 107:7805
243. Hodes G, Manassen J, Cahen D (1980) ibid. 102:5962
244. Hodes G, Manassen J, Cahen D (1981) J. Electrochem. Soc. 128:2329
245. Noufi R, Frank AJ, Nozik AJ (1981) J. Amer. Chem. Soc. 103:1849

Subject Index